Maria Rosecler Miranda Rossetto

Effect of gibberellic acid on carbohydrate metabolism in bananas

Maria Rosecler Miranda Rossetto

Effect of gibberellic acid on carbohydrate metabolism in bananas

Starch-Sucrose Metabolism

ScienciaScripts

Imprint

Cover image: www.ingimage.com

This book is a translation from the original published under ISBN 978-3-330-99626-7.

Publisher:
Sciencia Scripts
is a trademark of
Dodo Books Indian Ocean Ltd. and OmniScriptum S.R.L publishing group

120 High Road, East Finchley, London, N2 9ED, United Kingdom
Str. Armeneasca 28/1, office 1, Chisinau MD-2012, Republic of Moldova, Europe
Managing Directors: Ieva Konstantinova, Victoria Ursu
info@omniscriptum.com

Printed at: see last page
ISBN: 978-620-8-59087-1

SUMMARY

CHAPTER 1

INTRODUCTION

Ripening is a stage of development that is unique to fruit and begins after physiological maturity, the period of seed maturation. This event involves a large number of complex metabolic transformations. Although these reactions are largely degenerative in nature, the changes that occur seem to be part of the plant's natural senescence process. As a result, there are changes in colour, aroma, texture and flavour, making the fruit acceptable for consumption (Seymour et aL, 1996). These changes take place in a highly coordinated manner and are related to the activity of enzymes, with the participation of phytohormones in the initiation and development of these processes (Tucker & Grierson, 1984; Speirs & Brady, 1991).

One of the main biochemical events that takes place during fruit ripening is the mobilisation of reserve carbohydrates, which can be in the form of starch, organic acids or sucrose supplied by the plant itself, leading to the accumulation of sugars. The latter fits in with the case of non-climacteric fruit, where most of the formation and accumulation of the sugars responsible for sweetness depends on the mobilisation of sucrose from photosynthesis, sweetening very little after harvest (Seymour et aL, 1996).

In bananas, stored starch is the main source of reserves and is reduced during the climacteric period, from levels ranging from 12 to 20 %, depending on the banana cultivar, to less than 1% when the fruit is ripe. Concomitant with the hydrolysis of starch, the sucrose content increases approximately 12-fold and can reach 15% (Arêas & Lajolo, 1981; Mota et aL, 1997).

During this phase, the activity of sucrose-phosphate synthase (SPS), the enzyme responsible for sucrose synthesis in bananas, increases around 3-fold and remains high until fruit senescence, with a high correlation with sucrose synthesis and accumulation (Cordenunsi, 1989; Hubbard et aL, 1990; Mota, 1997). This correlation has also been observed in mango (Hubbard et aL, 1991) and kiwi (Macrae et aL, 1992). On the other hand, the activity of sucrose synthase (SS), an enzyme also linked to sucrose metabolism, begins to decline in the pre-climacteric period, at harvest time, reaching almost undetectable levels in the post-climacteric period. Since its maximum activity occurs during fruit formation, it can be deduced that the enzyme must be more involved in sucrose degradation than in synthesis, providing substrate (UDP-glucose or ADP-glucose) for cell wall and starch synthesis (Cordenunsi, 1989; Hubbard et aL, 1990; Nascimento et aL, 1997).

However, when the fruit ripens attached to the plant, these events take place at different speeds. Cordenunsi & Lajolo (1995) observed that fruit take longer to ripen and changes in SS and SPS activity occur more slowly, with the latter only reaching 50% of the maximum value obtained for harvested fruit. The most likely reason is that these events are related to the transmission of phytohormones from the plant to the fruit, which act as ripening inhibitors.

Fruit ripening is controlled by several classes of plant hormones, in a process that is still little known. Ethylene is by far the most studied hormone due to its importance as the factor responsible for fruit ripening and senescence. It is known that these processes depend, in part, on the increased sensitivity or decreased resistance of the fruit to the action of ethylene and also on the disappearance or inactivation of inhibitors, such as phytoregulators (Davies, 1995). Vendrell (1969) observed that when banana slices were infiltrated with auxin, the ripening of the fruit was delayed and the days of post-harvest conservation increased. Purgatto et al (2001) showed that banana slices infiltrated with the auxin indoleacetic acid (IAA) had their ripening delayed and some enzymes such as β-amylase had their activity profile altered during ripening. Gibberellins, auxins and cytokinins have been studied with particular interest because they function totally or partially as fruit senescence delayers. In general, gibberellins have the opposite effect to abscisic acid (ABA) in terms of increasing the synthesis of hydrolases, which are responsible for the degradative processes of leaf and fruit tissues (Cohen, 1996).

The aim of this work was to study the possible influence of gibberellic acid on the post-harvest sweetening process of bananas. We focussed on the content of soluble starch-sugars and some enzymes related to starch degradation and sucrose synthesis.

1.1 Fruit ripening: respiration and ethylene

In order to understand ripening, it is necessary to summarise the main events taking place in the fruit's cellular machinery. It can be said that most of the biochemical processes related to the ripening of climacteric fruit occur from the signalling promoted by ethylene and after the peak of respiration (Seymour et aL, 1996). Little is reported in the literature about the action of other phytohormones, such as auxin, gibberellins or cytokinins. Therefore, the fact that these substances act as senescence delayers is very attractive.

Respiration can be summarised as the oxidation of carbohydrates by the removal of hydrogen, which is channelled into the respiratory chain, generating a respiratory ratio between CO_2 and O_2. The main substrates used in this process are carbohydrates, lipids and proteins (Seymour et aL, 1996).

Bananas and other fruits classified as climacteric show a peak in respiration, described as respiratory climacteric, from which most of the events related to ripening take place. Some enzymes of the glycolytic pathway such as phosphofructokinase (EC 2.7.1.11) and fructose 1,6-biphosphatase (EC 3.1.3.11) have their activity increased during this period (Salminen and Young, 1975; Nair and Darak, 1981). The authors Bal et aL, (1991) suggest that the activation of pyruvate kinase (EC 2.7.1.40) and/or phosphoenolpyruvate carboxylase (EC 4.1.1.31) is the initial response of glycolysis in the climacteric. Thus, climacteric is characterised by a rapid rise in ethylene and CO_2. For bananas, the values for the increase in CO_2 observed were around 4 to 10 times, concomitant with the increase in ethylene production (Vendrell, 1970). In the Marmelo banana, for example, a 5-fold increase in CO_2 production was observed, and this respiratory activity was parallel to the rise in temperature, so variations in these levels can occur even within the same class of fruit (Chitarra, 1979).

The actual nature of the events that initiate the increase in ethylene biosynthesis has not been fully elucidated either. However, it has been observed that some messenger RNAs are transcribed and translated independently of ethylene. Adams & Yang (1979) suggested that ethylene in apple tissues is synthesised from methionine via S-adenosylmethionine (SAM) and a 4-carbon amino acid, 1-aminocyclopropane carboxylic acid (ACC):

L-Methionine→ SAM→ ACC→ Ethylene

In 1980, Yang proposed that this pathway was active in all higher plants. In this conversion, carbon 1 of methionine is converted into CO_2, carbon 2 into formic acid and carbons 3 and 4 into ethylene. The sulphur atom (S) is retained in the tissue.

The control of ethylene biosynthesis is regulated by its endogenous production which, in certain fruits, is caused by exposure of the tissue to exogenous ethylene. This considerable increase in endogenous ethylene production is called the autocatalytic effect, where ethylene itself stimulates the action of its synthesising enzymes: ACC synthase and ACC oxidase. It is assumed that the endogenous inhibitors existing in the pre-climacteric stage are destroyed by exposure to ethylene. Thus, ethylene production increases in the climacteric stage and the fruit ripens more quickly (Abeles et aL, 1992).

Two regulatory systems for ethylene biosynthesis are proposed: the first controls basal ethylene levels and stress-related production. It is present in both climacteric and nonclimacteric fruits. The second system regulates the autocatalytic production of ethylene and acts only in climacteric fruit.

A gene family of ACC synthase and ACC oxidase enzymes are involved in this process, along with receptor proteins, which form a complex signal transduction mechanism within the tissue, acting on the genes of these enzymes (McKeon et al., *In:* Davies, 1995). Both the formation of ACC and its subsequent conversion into ethylene are inhibited by polyamines, as both utilise a common precursor (SAM) and also a common by-product (MTA) (Page & Willians-Ashman, 1969; Sagee et al., 1980; Biondi et al., 1990). SAM can be used in the biosynthesis of polyamines or ethylene or both, depending on the conditions that regulate its metabolic pathway, although the function of the metabolism of these substances in the plant is different (Even-Chen et al, 1982).

The process regulated by ethylene in fruit ripening has been studied extensively with the aim of prolonging storage time and improving fruit quality. Through research into the enzymes responsible for synthesising this hormone, it has already been possible to block the gene transcription of these enzymes in tomatoes and melons using the antisense RNA technique (Hamilton et al., 1990; Oeller et al., 1991). However, it has been observed that changes in fruit metabolism do not appear to be related to the isolated action of a single phytohormone, but rather to be the consequence of a hormonal balance (Gillaspy et al., 1993; Cordenunsi & Lajolo, 1995). The most important effects of the ethylene regulation mechanism in fruit can be seen in the diagram in figure 1.

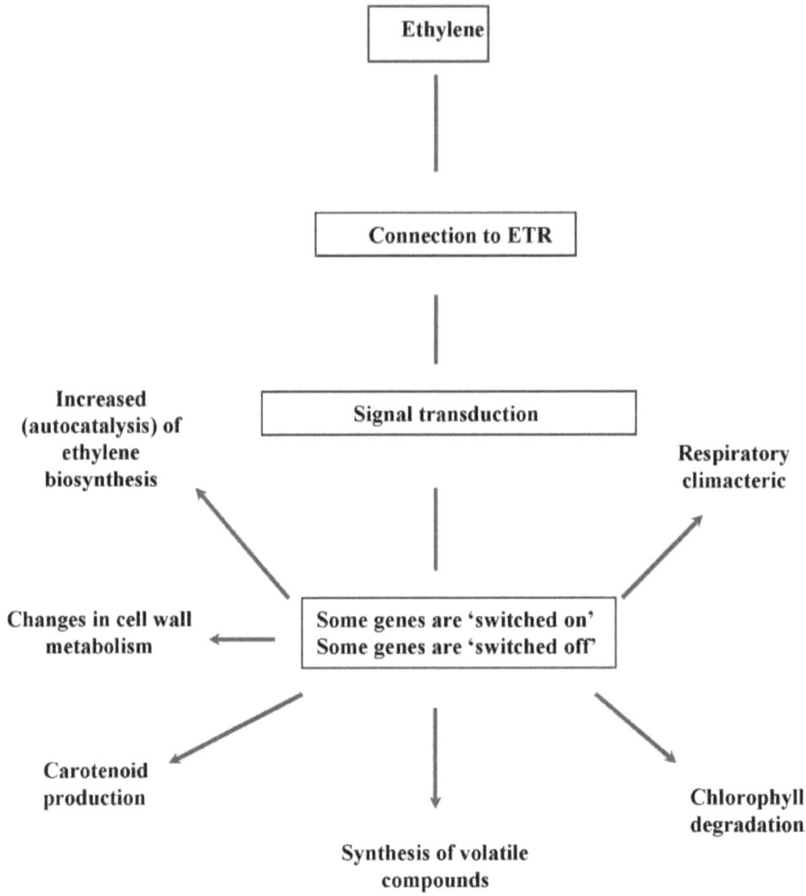

Figura 1: Schematic representation of the mechanism of action of ethylene in climacteric fruit.

Notification:

ETR (protein): ethylene receptor that triggers a cascade of bio-chemical events in the tissue through signal transduction.

1.2 Starch degradation and sucrose synthesis in bananas

Starch-sucrose metabolism is one of the main events observed during banana ripening. The degradation of starch (around 20 per cent) and the accumulation of soluble sugars, mainly sucrose (around 15 per cent), begins in the climacteric period of the banana, concomitantly with the increase in respiration and ethylene.

Starch is a polysaccharide unique to plants in that it consists of granules that can be isolated. They are organised within specific plastids (chromoplasts): the amyloplasts. After the rupture and dissolution of these granules, most of the starch can be fractionated into two macromolecules, which are glucose homopolysaccharides, defined as amylose and amylopectin. They have different properties in terms of solubility, type of bonds and chain size, iodine complexing power, degree of retrogradation and molecular weight. Analytical and microscopic techniques have made it possible to see the different levels of organisation of these granules (Gallant et aL, 1997).

In intact banana cells, starch granules are heterogeneous, differing in size and shape depending on the cultivar. They appear irregular under the optical microscope, with the small ones being spherical and the large ones oval or elongated (Lii et aL, 1982; Garcia & Lajolo, 1988; Motta, 1997). It is thought that there are two ways of initiating starch degradation: endo- and exocorrosion of the granules, but both can occur together. Garcia & Lajolo (1988) observed that as the banana ripened, the number of small granules increased to the detriment of the large ones, which were preferentially degraded. They also observed that the surface of the granules of green bananas was smooth, while the granules of ripe bananas showed striations, hypothesising that the attack by the enzymes occurred through a process of exocorrosion.

These striations can be seen in the granules of various cultivars, suggesting that the action of amylases begins well before the onset of climacteric (Mota, 1997).

Starch degradation is carried out through a complex enzymatic process (figure 2). The enzymes involved in this metabolism can coexist in multiple forms and be located outside the plastid, which makes it difficult to establish the metabolic pathway (Steup, 1988).

The α-amylase (EC 3.2.1.1) randomly hydrolyses the α-1,4 bonds of amylose and amylopectin, producing a mixture of oligosaccharides (dextrins). Garcia and Lajolo (1988) detected the activity of 3 α-amylases, 4 β-amylases and α-1,4 and α- 1,6-glucosidases during banana ripening, and observed that the activity of these enzymes increased in the climacteric phase, hypothesising that they were directly related to starch degradation. It is already known that banana α-amylase appears to have a molecular weight of 66 kDa, while most endoamylases from other plants have subunits ranging from 42 to 45 kDa (Witt and Sauter, 1996). It is also known that all forms of α-amylases have one Ca^{2+} atom per molecule, which confers stability to the molecule (Whitaker, 1994).

β-amylase (EC 3.2.1.2) acts only on the penultimate bond of the non-reducing end of amylose and amylopectin, releasing only maltose. In bananas, the absence of maltose as a common sugar suggests a synchronised action of β-amylase and α-(1,4)-glucosidase (Garcia, 1983), and it has been proposed that α- and β-amylases act together to initiate the breakdown of starch, since β-amylase does not show activity on starch granules that have not previously been digested by a-amylase (Dunn, 1974). Most P-amylases are monomeric, with the exception of sweet potato, which has been shown to be a 215 kDa tetramer, with subunits around 55 to 64 kDa (Monroe & Preiss, 1990; Wong, 1995). Banana β-amylase appears to be a dimer composed of 56 kDa subunits and has a pronounced increase in activity and expression during the starch degradation phase (Purgatto et al., 2001).

The degradation products of α- and β-amylases that have α-1,6 glycosidic bonds are metabolised by desramifying enzymes (EC 3.2.1.41; EC 3.2.1.10 and EC 3.2.1.68) to maltotriose, maltose and oligosaccharides. Maltose can be hydrolysed to glucose by the action of α-glucosidases (EC 3.2.1.20). However, no debranching enzyme has yet been identified in fruit, but glycosidases have already been studied in banana, some isoforms of α-glucosidases, in soluble form, attached to the cell wall have been purified and characterised, the soluble form of this enzyme is a typical maltase. The activity of this enzyme has been shown to be stable in the climacteric phase, acting at an optimum pH of 4.5, with a molecular weight of around 70KDa (Konishi et al., 2001).

Phosphorylases (EC 2.4.1.1), together with amylases and glycosidases, take part in starch metabolism in bananas, both by producing glucose-1-phosphate from the α-1,4 bond of the non-reducing terminal in the presence of Pi, and in the opposite direction, releasing inorganic phosphorus (Arêas & Lajolo, 1981). Two forms of the enzyme were observed in bananas with characteristics similar to the cytosolic and plastidial forms described in the literature for other plant tissues (Mota, 2001, in press). The plastidial form of the enzyme seems to act on the starch granule and the cytosolic form, as it is distant from the substrate, its function is not known for sure (Buchner et al., 1996). Very little is known about the regulation of the forms of this enzyme, however, it is known that starch phosphorylase is inhibited by heavy metals, such as Hg^{2+}, Ag^{+} , Cu^{2+}, $Fe^{(2+)}$, $Zn^{(2+)}$), and in banana leaves it has been observed that adenosine triphosphate (ATP) also inhibits its activity. The optimum pH range found for the enzyme in both animals and plants was around 6.0 to 7.0, and it is rapidly inactivated in the pH range below 4.9 (Nighojkar & Kumar, 1997).

The end products of starch degradation, glucose and glucose-1P, can be converted to glucose-6-phosphate and fructose-6P by the action of hexokinase (EC 2.7.1.1) and phosphoglucomutase (EC 5.4.2.5), respectively. The products of the action of these enzymes will be used to synthesise sucrose, and glucose can be used to start the respiration cycle in glycolysis. Their utilisation occurs especially in the cytoplasm (Tucker, 1993).

The enzyme responsible for the interconversion of glucose-6P into fructose-6P is phosphoglucose isomerase (PGI) (D-glucose-6-phosphate-ketol-isomerase) (EC 5.3.1.9). Its action ensures the release of fructose-6P, which is the substrate for sucrose phosphate synthase (SPS). In parallel, the enzymes phosphoglycomutase and UDPG pyrophosphorylase are activated to produce UDPG, a nucleotide sugar that is used to form sucrose (Ap Rees, 1980). According to Kruckeberg et al. (1991), PGI occurs in the plastids and cytosol. PGI in the plastid is essential for starch synthesis, while PGI in the cytosol acts in the conversion of triose-phosphates to sucrose. Mota (1997) found that the peak of PGI activity coincided with the accumulation of sucrose in bananas.

Sucrose synthesis is basically mediated by two enzymes that appear to be homogeneously localised throughout the banana pulp (Bassinello et al., 1999), 1999), through two alternative pathways: the first through the action of sucrose phosphate synthase (SPS), producing sucrose-6P from UDP- Glucose and fructose-6P: (UDPG + F-6P→ sucrose-6P + UDP), followed by sucrose phosphatase (EC 3.1.3.24), which removes the phosphate group from the molecule, generating sucrose. The second would be through the action of sucrose synthase (SS), which produces sucrose from UDP-glucose and fructose: (UDPG + fructose←→ ,sucrose + UDP). In bananas, it has been observed that the SS has a much lower activity once it is disconnected from the plant, being more involved in the fruit's development phase. Banana SS is a tetramer, found in two isoforms that differ in their isoelectric point (4.59 and 4.77). Each subunit of the enzyme has a molecular weight of around 90 kDa and control of its activity is based on the level of SS protein and post-translational modification, which involves gene transcription of the enzyme (Nascimento et al., 1997; Cordenunsi & Lajolo, 1995; Nascimento et al., 2000). Sucrose can also be hydrolysed by the action of invertases (EC 3.2.1.26), and directed towards the *de novo* synthesis of glucose and fructose (Leloir & Cardini, 1955).

The key enzyme for sucrose synthesis in bananas is SPS, since its activity/expression increases along with starch degradation and sucrose synthesis (Cordenunsi & Lajolo, 1995; Nascimento et al., 1997). SPS activity in higher plants seems to be regulated by two distinct mechanisms. The first seems to be related to allosteric regulation through the balance between activators and inhibitors of enzyme activity. Similar to banana, it was observed in spinach leaves that SPS was activated by glucose-6P, by increasing the speed of the reaction and the affinity of the enzyme for fructose-6P (Doehlert & Huber, 1983 (a, b); Nascimento et al., 1997). However, it was found that this activation is subject to inhibition by Pi (inorganic phosphate) which can act as a partial competitive inhibitor of the enzyme.

A second proposed regulatory mechanism is that the change in the enzyme's activity can occur due to alterations in its kinetic properties, i.e. the enzyme can undergo a covalent modification by reversible phosphorylation or alter the levels of the SPS protein. It has been proposed that banana SPS is activated by the phosphorylation mechanism (Hubbard et al., 1990) and by an increase in protein (Nascimento et al., 1997). The molecular weight of banana SPS is around 420 kDa, with subunits of 116 kDa, values close to those found for spinach (Sonewald et al., 1993), maize (Bruneu, et al., 1991) and beetroot (Hesse et al., 1995).

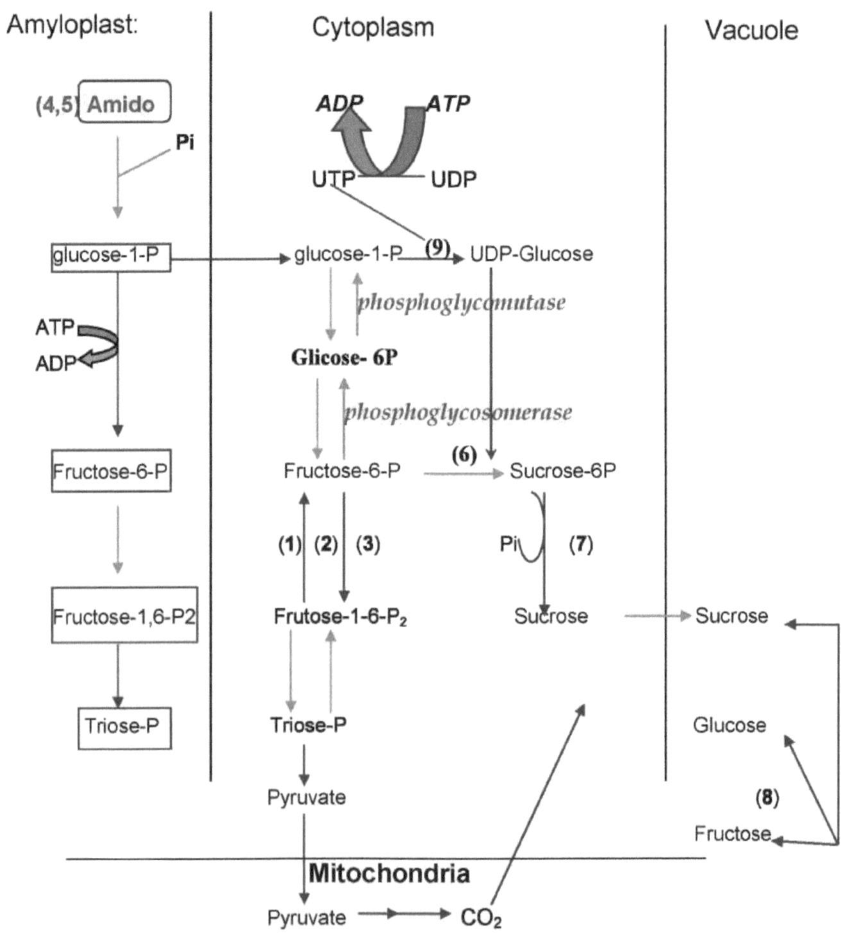

Figura 2: Probable metabolic pathway of starch-sucrose transformation in higher plants, partly proposed by Hubbard et al.(1990), where: (1):Fructose 1,6 bisphosphatase; (2): ATP phosphofructokinase;(3): Pyrophosphate-dependent phosphofructokinase; **(4)**

phosphorylase; **(5)** α- and β-amylases and α-1,6 glycosidase; **(6)** Sucrose phosphate synthase; **(7):** Sucrose phosphate phosphatase; (8) Invertase; (9) UDP pyrophosphorylase.

1.3 Mechanism of action and effect of gibberellic acid on fruit

Phytohormones are organic substances that occur naturally in plants and influence physiological processes in low concentrations. *Gibberellins* were discovered by the Japanese Kurosawa in 1926, when he was studying a disease caused by the fungus *Gibberella fujikurai*, which caused abnormal elongation of the rice stem (Fosket, 1994). They are a family of plant hormones with more than 100 compounds already identified, few of which are active. They all have a common "ent- *giberellane*" ring structure of tetracyclic diterpenoid acids, differing in the positions and spatial rearrangements of their hydroxyls. In their biosynthesis they use the common precursor of all diterpenoids: geranyl geranyl diphosphate (GGPP; C=20), which is converted to kaurenoic acids by the action of ent-kaurene synthase and specific kinases, giving rise to the gibberellin GA_{12}, which is the precursor of all other gibberellins, including gibberellic acid (Davies 1995; Garcia-Martínez & Hedden, 1997).

The common synthesis pathway is directed towards the formation of carotenoids by the action of phytoene synthase, or towards the formation of sterols, essential substances in the cell membrane. Gibberellins are substances that are produced by the secondary metabolism of plants and can be biosynthesised by fungi (Fernádez-Martín et al., 2000). They are biosynthesised by the fungus *Gibberella fujikurai*, from which they were discovered (Davies, 1995). It has been observed that *Fusarium moniliforme* can also produce them (Fernández-Martín et al., 2000).

The whole plant is capable of biosynthesising gibberellins, but the main production is in the seeds, where most of the hormones are synthesised in considerable (low) quantities (Davies, 1995). Inside the tissue, the level of gibberellins is controlled by conjugation with glucose molecules, esters and ethers. These substances bind, momentarily inactivating the endogenous gibberellin. However, when the conjugation is made with amino acids and proteins (less common), there is almost definitive inactivation of these substances, as it is much more difficult to break these groups (Martínez & Hedden, 1997). In *Arabdopsis* dwarf mutants, it has been observed that gibberellins are catabolised by GA_{20}-oxidase. Gibberellins can replace the dwarf phenotype in some plants, as they act on the tissue by increasing cell size and also promote tissue elongation. It was observed that even though mutant plants contained a high amount of gibberellins, this was not enough to replace dwarfism when the enzyme GA_{20}-oxidase was present (Harberd et al., 1998).

It was observed that the ovaries of different parthenocarpic fruits contained a much greater amount of endogenous gibberellin when compared to normal fruits, in addition to being seedless, these fruits were firmer and had a higher soluble solids content (Gillaspy et al., 1993). The same authors, when studying the hormonal balance of tomatoes, showed that gibberellins were only present in the first three main phases of fruit development. However, when the fruit reached physiological maturity, the phytohormones present were only auxin and ethylene, noting that the endogenous gibberellin had disappeared. In the seed development phase, cytokinins and abscisic acid were the most important, while auxin and gibberellin were less important. It is therefore surmised that endogenous gibberellin could contribute to the increase in auxins, and as observed by Henry et al. (1967), this auxin would be responsible for helping to maintain the juvenility of the tissue (Gillaspy et al., 1993).

Some active gibberellins act alone, or together with auxins, with a senescence-delaying effect. Gibberellins keep the texture of the fruit firm, promote changes in colour by delaying the degradation of chlorophyll and the accumulation of carotenoids and anthocyanins. Gibberellic acid is one of the low-cost active forms widely used in agriculture. A schematic of part of the gibberellic acid biosynthesis pathway can be seen in figure 3.

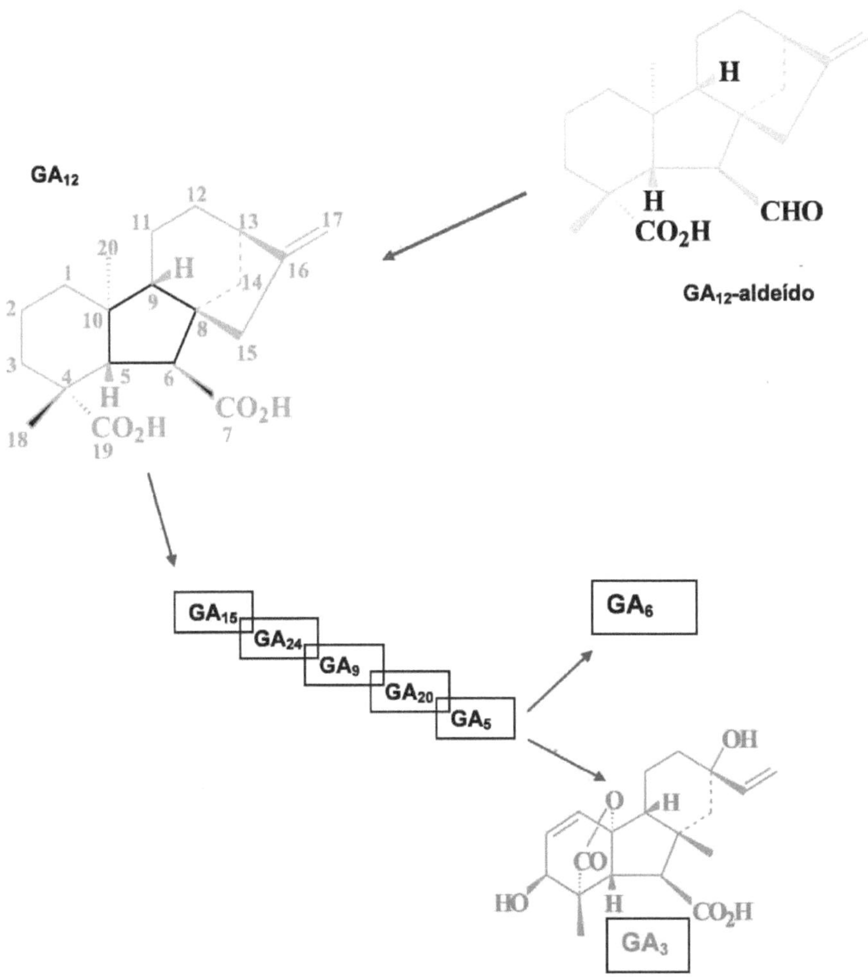

Figura 3: Partial schematic representation of the gibberellic acid biosynthesis pathway (Davies, 1995)

Citrus fruits treated with GA_3 can be kept on the tree and the harvest postponed for several weeks, as the degradation of chlorophyll is delayed. This means that young peel tissues offer better protection against physiological or pathological disorders, mainly aimed at protecting against fruit flies (Biale, 1978; Abdel - Gawad & Romani, 1974; Marur et al., 1999). It has been shown, both in experiments with slices and whole strawberry fruit, that exogenous application of GA_3 induces a significant decrease in the respiration rate, decreases the synthesis of anthocyanins and the degradation of chlorophyll, decreasing the activity of related enzymes such as chlorophyllase and peroxidases (Martínez et al., 1996).

The application of gibberellic acid-based products to citrus fruits is a common practice and has an important effect not only by delaying ripening, but also by making it uniform. These are determining factors for scheduling the harvest and obtaining better prices for the product on the market (Marur et al., 1999).

Winegrowers in Rio Grande do Sul are using products based on gibberellic acid and GA_1 to promote elongation and cell enlargement of the berries (Zabadal & Dittmer, 2000). As a result, they are spaced further apart, preventing the formation of microclimates favourable to fungi, as well as promoting greater firmness of the fruit near the pedicel, which prevents the berries from falling prematurely (Peres & Morales, 1999).

Gibberellic acid seems to play a particularly important role as a regulator of hydrolytic enzymes. It has been observed that the balance between the levels of GA and endogenous ABA controls the gene expression of a-amylases in germinating seeds and, in the case of barley, increases the transcription of genes and the activity of a-amylases (Davies, 1997; Loreti, 1998).

Based on studies of increased gene transcription of α-amylases in cereals, some authors have proposed a signal transduction pathway for gibberellins. It is interesting to note that only some cells develop the ability to increase gene expression, as was the case with cells in the aleurone layer of cereals. This cellular competence is acquired throughout development and does not seem to depend on hormones (Fosket, 1994).

The transduction pathway in the aleurone cell layer has been proposed by some authors, who assume that the receptor, as yet unidentified, is located in the plasma membrane, and report that the G-protein (heterotrimeric) is a good candidate to be a signal receptor. The secondary messengers, calcium, calmodulin, inositol triphosphate, diacylglycerol, cyclic GMP, activate kinases inside the cells, promoting the phosphorylation of proteins and a favourable change in cytosolic pH. The messengers are able to conduct the signal to its end in the cytosol, to act in the nucleus and induce the GaMyb gene transcription factor, increasing the synthesis of α-amylase mRNA (Gubler et al., 1995; Rogers & Rogers, 1999; Thornton et al., 1999). The kinetics of the expression of the GAMyb gene indicate that it is a factor that regulates the gene expression of α-amylase signalled by gibberellin. It is hypothesised that this transcription factor is part of the response mechanism to the induction of these enzymes (Gubler et al., 1995).

A reduction in the activity of amylases and peroxidases, induced by GA_3, was observed in mango. Post-harvest application of the hormone in this model significantly delayed ripening by slowing down the degradation of ascorbic acid and chlorophyll in the peel, as well as observing a reduction in the activity of α-amylases during storage (Khader et al., 1988; Khader, 1992). Gibberellic acid has been shown to inhibit respiratory weathering in peaches (Abdel - Gawad and Romani, 1974) and tomatoes (Babbitt et al., 1973). These observations confirm others which suggest that tissue sensitivity to ethylene can be controlled by other classes of phytohormone, blocking or increasing ethylene synthesis during abscission, ripening and senescence (Abeles et al., 1992)Little is known about the influence of hormones on banana ripening. In 1970, Vendrell, working with slices and whole banana fruits, vacuum infiltrated different concentrations of gibberellic acid and observed that the treated slices had normal ethylene production, soluble solids content, pulp softening and the presence of volatiles.

Therefore, he observed that there was a delay in chlorophyll degradation and peak respiration for around five days when 0.1 mM of gibberellic acid was infiltrated into the whole fruit. The author, along with others, hypothesised that gibberellic acid could act on banana tissue in two ways: 1) in the hormonal system, contributing to an increase in auxin synthesis and 2) possibly in the enzymatic system (Henry et aL, 1967).

CHAPTER 2

OBJECTIVES

This work had the following objectives:

1) To study the possible involvement of gibberellins, particularly gibberellic acid, in delaying banana ripening.
2) To verify the influence of gibberellic acid on amide-sucrose metabolism and the activity/expression of some related enzymes.

CHAPTER 3

MATERIAL AND METHODS

2.1 Material

The bananas *(Musa acuminata* (AAA) Cv. Nanicão) used in the experiments were obtained from the Federal University of Viçosa (MG) and CEASA (SP). In the first case, the bananas were transported to the laboratory by bus and packed in cardboard boxes. All the bananas were bought at approximately 110 days post anthesis (dpa), at the point of commercial harvest. The bananas from CEASA, coming from Vale do Ribeira (SP), were transported in wooden crates by lorry. The fruit used in the trials was not subjected to ethylene acclimatisation.

2.2 Methods

3.2.11 Gibberellic acid infiltration in banana slices

The fruit was washed and then immersed in a 1% sodium hypochlorite solution for three minutes to reduce surface contamination. They were then dried with absorbent paper and cut down the centre with a stainless steel blade into slices approximately 6 mm thick and sorted by weight.

All the material used in the infiltrations was autoclaved or sterilised with 70% ethanol. The infiltrations were carried out under aseptic conditions in a laminar flow hood, submerging the treated slices in a 120 mM mannitol solution with 0.1 mM gibberellic acid (Sigma), while the control slices were submerged in a 120 mM mannitol solution. The mannitol was added to maintain the cell's osmolarity in order to reduce tissue stress. The infiltrations were controlled by weight gain, which was considered satisfactory at between 6 and 10 per cent. Ripening was monitored by measuring respiration (CO_2) and ethylene in the banana slices, as well as the colour of the peel and the aroma of ripe fruit. Slices contaminated with fungi were discarded.

Based on the respiration curve, samples were selected and collected at three-day intervals until the climacteric peak, from which point collections were made every day until the control slices had fully ripened. Each triplicate slice consisted of a sample that was pulverised and frozen in liquid nitrogen and stored in an ultrafreezer at -80°C until use.

3.2.1.1 *First experiment:*

The first trial used bananas from CEASA (SP). The slices, made and infiltrated as described above, were placed in desiccators (figure 5) and kept in a flow (3.0L/h) of humid air (85-90% RH) free of CO2, by passing them through a 20% KOH solution in a respirometer (figure 4). The bottom of the desiccators contained enough water to maintain relative humidity at around 90% and the slices were isolated from direct contact with the water using glass beads. Respiration measurements were carried out on slices separated especially for this purpose and packed in triplicate jars (volume = 1 L), following the same procedures as the slices used for collection. Ripening was monitored by daily measurements of CO2, peel colour, ripe banana aroma and starch analysis. The aim of this experiment was to optimise working conditions for the two definitive trials.

3.2.1.2 *Second experiment:*

Banana slices of approximately 110 dpa from the University of Viçosa (MG) were used. Ripening was monitored using gas chromatography, by measuring respiration and ethylene, as well as peel colour and the aroma of ripe bananas. The test was carried out again in a respirometer set up as described in the first experiment, illustrated in figure 4. However, the volume of the respiration flasks was optimised to 500 ml, where part of it was again occupied with glass beads to isolate the slices from the bottom of the flask, leaving a useful volume of 400 ml, used in the calculations. Air humidity was maintained at around 85-90% by adding water to the bottom of the flasks and desiccators (figure 5). The flow rate was reduced to 1 L/h in order to induce a small accumulation of ethylene and promote normal ripening of the fruit, as it was observed in the control slices from the first experiment that the fruit took a long time to ripen, perhaps the basal ethylene was not enough to promote normal ripening of these fruits. The slices to be analysed in this second experiment were placed in desiccators under the same conditions as the breathing vials. Slices contaminated with fungi were discarded again. Glass spheres were placed back in the bottom of the jars and desiccators, with the slices placed on top of them to avoid direct contact with water. The slices were collected again from the desiccators every 3 days until the climacteric peak, from which point they were collected every day. The samples were then frozen and pulverised in liquid nitrogen and stored in a -80°C ultrafreezer.

3.2.1.3 *Third experiment*

The experimental conditions used for this trial were exactly the same as for the second one, with the aim of obtaining more samples to continue the work. The bananas were purchased from CEASA at approximately 110 dpa (commercial harvest point). Ripening was again monitored by GC, peel colour and the aroma of the slices. Collections were alternated until the day of the ethylene peak, after which they were repeated every day until the control slices were completely ripe.

Figure 4: Respirometer

Figure 5: Collection desiccators (right side) and breathing bottles (vol=1L) (for collecting CO_2 and ethylene gases) (left)

3.2.2 Starch analysis

The starch content was determined enzymatically using the method described by Arêas & Lajolo (1981). The starch was extracted with 0.5N sodium hydroxide, precipitated with 80% ethanol, hydrolysed with amyloglucosidase and the glucose released was determined using the glucose oxidase/peroxidase/ABTS system, according to the method of Bergmeyer (1974).

3.2.3 Determination of soluble sugars

The soluble sugars (glucose, fructose and sucrose) were extracted three times with 80% ethanol at 80°C, the supernatants were combined, the ethanol evaporated using a vacuum system at low temperatures (speed vac) and the volume reconstituted with water.

Sugars were estimated by HPLC ion exchange chromatography using a DIONEX model DX500 chromatograph coupled with a pulse amperometric detector, model ED40. The column used was CARBOPAC PA 1 (4x250 mm). The mobile phase consisted of 100% 18 mM NaOH at a flow rate of 1.0 mL/min.

3.2.4 *Preparation of enzyme extracts*

The enzyme extracts for measuring SS and SPS activities were made following the methodology optimised by Cordenunsi (1989), where around 0.5g of banana pulp was homogenised in 2.0 ml of extracting medium in a Brinkman homogeniser (Turrax type) for around 30 seconds at medium speed. The homogenates were centrifuged at 7650xg for 5 minutes and the supernatants desalinated in Sephadex mini-columns (Pharmacia). The extraction solution consisted of 100 mM Tris-HCl pH 8.0 buffer containing 20 mM EDTA; 20 mM PVP 40000 1% cysteine and 1 mM benzamidine.

Amylase extraction was carried out according to Garcia (1988), where around 0.5g of frozen banana sample was homogenised in 1.25 ml of extractant solution in a Turrax for 30 seconds at medium speed. The homogenates were centrifuged at 12,000 x g for 30 minutes and the activities were measured in the supernatant. The extraction medium consisted of 100 mM Tris-HCl buffer pH 7.0 containing 20 mM cysteine, 1 mM benzamidine and 1% PVP 40000. The Tris-HCl buffer was also used because it inhibits glycosidases, thus preventing these enzymes from interfering with amylase activity. The β-amylase extraction was carried out under the same conditions, but the buffer was changed to Hepes-KOH 50 mM pH 7.0, due to the specific method for the activity that acts on α-glucosidases enzymes, the extract was centrifuged at 8000 rpm and the activity measured in the supernatant, in the crude extract.

The phosphorylases were extracted according to the method described by Arêas and Lajolo (1981), where around 0.5 g of banana was homogenised in 3.0 ml of extractant solution in Turrax for around 30 seconds at medium speed and then centrifuged at 8000 rpm for 50 minutes. Phosphorylase activity was measured in the supernatant. The extraction medium consisted of Hepes-KOH buffer 50 mM pH 7.5 containing EDTA 20 mM, cysteine 20 mM, PVP 40000 1%.

3.2.5 *Activity of sucrose-phosphate synthase (EC 2.41.14) and sucrose synthase (EC 2.4.1.13)*

SPS activity was measured in the direction of sucrose synthesis. The activity medium was composed of 25 µl of the extract and 176 mM Hepes-KOH buffer pH 7.5, containing 33 mM NaF, 50 mM fructose-6-P, 150 mM glucose-6-P, 100 mM UDPG, 500 mM MgCh, in a final volume of 50 µl. The activity medium was incubated at 30°C. The reaction was stopped by adding 200µl of 0.1 M NaOH and boiling for 10 minutes. The sucrose formed was determined using the thiobarbituric acid method (Percheron, 1962).

SS activity was determined in the sense of sucrose hydrolysis, where 25 p.1 of the enzyme extract was incubated in a medium composed of 330 mM Tris-HCl buffer pH 7.6 containing 33 mM NaF, 100 mM fructose, 50 mM UDPG and 500 mM MgCfe in a final volume of 50 µl. The activity medium was incubated at 30°C. The reaction was stopped by adding 200µl of 0.1 M NaOH and boiling for 10 minutes. The sucrose formed was determined using the thiobarbituric acid method (Percheron, 1962).

3.2.7 Total amylase activity

The activity of the amylases was measured according to the method of Caraway (1959), which measures total hydrolytic activity. A 0.01 N starch solution was used as the substrate. The activity medium contained 100 µl of enzyme extract and 1 ml of starch in water. After incubating the sample/substrate at 37°C for 30 minutes, the reaction was stopped with 1 ml of 0.01 N iodine solution in strongly acidic medium (HCl), 4 ml of water was added to the solution to enable the reading. The amylase activity was estimated using the colour formed by the iodide solution and the remaining starch at 660 nm. One unit (1U) of amylase activity was arbitrarily defined as the amount of enzyme needed to hydrolyse 1 mg of starch in 15 minutes at 37°C in 1 ml of sample.

3.2.8 Activity of cx-amiassss (1,4 α-D-alianoo glycanohydrolase, EC3.2.1.1)

The α-amylase activity was measured according to the method of Hagenimana et al. (1992) specific for a-amylase, using azure amylose (Sigma) as a substrate (36 mg amylose-azure / ml) in 0.02 M Tris-HCl buffer pH 7.0, in a 1:1 ratio of enzyme extract/substrate at 37°C for 1 hour. The reaction was stopped with 160 µl of 18% (v/v) acetic acid and the samples centrifuged at 8000 x g for 15 minutes. The supernatant was read at 595 nm. One unit (1U) of a-amylase activity was defined as the amount of enzyme capable of releasing 1µg of soluble dye (remazol brilliant blue - RBB) per gram of banana pulp in 1 minute of reaction under the conditions described, based on the RBB standard curve (Sigma).

3.2.9 *ACTIVITY OF --AMILSSE*

The activity of β-amylase was measured according to McCleary and Codd (1989), using a specific substrate, ρ-nitrophenyl-maltopentaoside (PNPG5, 5 mM). The reaction takes place in the presence of the enzyme α-glucosidase, which releases the compound ρ-nitrophenol which develops a yellow colour at high pH. Betamyl (Megazyme) was used, which contains a mixture of α-glucosidase (1000 U/vial) and PNPG5 (47.5 mg/vial). The reaction was carried out in a medium containing 50 µl of the crude extract and 50 µl of the substrate (a-glucosidase 20 U + PNPG5 5 mM), at 30°C for 20 minutes. The reaction was stopped with 750 µl Trizma-base 1% (pH>10). Reading, according to the methodology, was carried out at 410 nm. One unit (1U) of β-amylase activity was defined as the amount of enzyme capable of releasing 1 pmol of ρ-nitrophenol per minute. The calculations were based on the standard curve of the ρ-nitrophenol dye.

3.2.10 *PHOSPHORYLASE ACTIVITY (α-1,4-D-GLUCAN: ORTHOPHOSPHATE-GLIOOSYL TRANSFERASE,* EC *2.4.1.1.):*

Activity can be measured in the direction of synthesis (phosphate release) or starch degradation (glucose-1-P release), as described by Arêas and Lajolo (1981). Experimental data in our laboratory has shown that there are no significant differences in the activity profile between the synthesis and degradation starch by phosphorylases. In this work, phosphorylase activity was measured in the direction of starch synthesis, as it is a colourimetric method that is quicker and cheaper than the enzymatic method, according to Lowry and Lopez (1946). The reaction was carried out in a medium containing 300 µl of 50 mM Mes-NaOH buffer pH 6.0, 50 µl of 50 mM NaF, 50 µl of a starch solution (dissolved in the same reaction buffer) and 50 µl of crude extract, in a final volume of 450 µl. The samples were incubated at 30°C for 10 minutes, together with 150 mM glucose 1-phosphate. The reaction was stopped with an ice-cold solution of 5% trichloroacetic acid (TCA), precipitating the proteins. Then 4 volumes of 0.1 M sodium acetate were added to raise the pH to 4.0.

The phosphate formed was determined by the action of a reducing agent for phosphorus, i.e. 0.1 volume (500 µl) of ascorbic acid (1%), followed by the addition of 0.1 volume (500 µl) of ammonium molybdate (1%), the samples were incubated again at 37 °C for 15 minutes. The phosphomolybdate formed was determined by spectrophotometry at

700 nm, where 1U of enzyme activity corresponds to 1μ mol of $PO4^{3'}$ released per minute.

3.2.11 *Protein Quantification*

Protein determination was carried out on crude extract according to the Bradford method (1976) to obtain the specific activity of the enzymes under study. The protein bovine serum albumin (BSA) was used as a standard. In the specific case of protein quantification for electrophoresis, the method of Lowry et al. (1951) was used.

3.2.12 CO2 and ethylene measurements

CO_2 and ethylene were measured using gas chromatography coupled with a flame ionisation detector for ethylene (FID) and a thermal conductivity detector for CO2 (TCD). The column used to measure the gases was HP-PLOT, which has a length of 30 m, a film thickness of 40.0 μm and an internal diameter of 0.53 mm. The gases were injected manually using a GAS *Tigth* syringe suitable for injecting gases. The samples were taken from sealed vials and left in the flow of the respirometer for one hour until all the air had been replaced by the gases released from the slices in the vials, which contained an average of 60g of banana. The carrier gas used for both analyses was helium.

3.2.12.1 Chromatographic conditions for CO2:

Isothermal temperature (oven): 30°C
Injector and detector temperature: 250°C
Flow: 4 ml/min
Injection mode: *split, split* ratio 50:1
Injection volume: 1 ml

3.2.12.2 Chromatographic conditions for ethylene:

Isothermal temperature (oven): 30°C
Injector temperature: 200°C, detector: 250°C
Flow: 1 ml/min
Injection mode: *pulsed split,* Rate: 12.5:1
Injection volume: 5 ml

3.2.12.3 **CO2 and** ethylene calculations

The CO2 and ethylene measurements were calculated based on the free volume of the jar, taking into account the mass of banana contained in the breathing jars and the air flow, which are given in mg of CO2.Kg PF-1 .h-1 and µl of ethylene. $Kg.PF^{-1}.h^{-1}$, respectively.

3.2.13 Extraction of total protein for Western Blotting

Around 0.5g of banana was homogenised in 2.5 ml of extraction medium, consisting of 0.25 M Tris-HCl buffer (pH 8.4), 0.2 M glycine, 0.4% SDS (w/v) and 10% β-mercaptoethanol (v/v), according to Laemmli (1970). The homogenate was centrifuged at 12,000 x g for 20 min.

The protein in the supernatant was precipitated with 12% TCA on ice overnight. The extract was centrifuged again at 12,000 x g for 20 min and the supernatant discarded. The precipitate was washed and centrifuged in an ice-cold solution of ethanol-ethyl ether-chloroform (2:2:1), then acetone and finally ethyl ether.

After evaporating the ether, the samples were resuspended in Laemmli buffer (1970) and boiled for at least 10 minutes.

3.2.14 Polyacrylamide gel protein electrophoresis

Electrophoresis was carried out on polyacrylamide gels (7.5%) under dissociating conditions (SDS-PAGE), according to Laemmli (1970), in 25 mM tris buffer; 192 mM glycine (pH 8.3), containing 0.1% SDS. Two gels, one containing control samples and the other treated, were revealed in Comassie Blue-R solution to check the total protein. 20 p.g of protein was used in each gel.

3.2.15 Western Blotting

Four gels made under the conditions of item 3.2.14 were used for Western blotting. Two were used for phosphorylases, one with control samples and the other with treated samples, with 5 µg of protein in each of the gels. Another two were used for SPS (control and treated) containing 10 µg of protein.

The gels were pressed onto nitrocellulose membranes in 25 mM tris 192 mM glycine buffer (pH 8.3). The voltage was set at 30 V and the transfer was carried out overnight at 4°C. The membranes were stained for protein with 0.2% Ponceau-S/ 3% TCA to check that the bands had been transferred and to mark the standards.

The membranes were blocked by incubating them for one hour in a solution containing 50 mM Tris-HCl, 150 mM NaCl and 5% skimmed milk at pH 7.5. Next, rabbit antiserum against SPS (Nascimento, 1997) and rabbit antiserum against phosphorylase (Mota, 2001) were added to the membranes at dilutions of 1:1000 and 1:500, respectively. The membranes were kept stirring for two hours at room temperature. The membranes were then washed three times in a 50 mM Tris-HCl solution pH 7.5, plus 150 mM NaCl (TBS buffer). After washing, the membrane was again incubated with the conjugated antibody (horseradish peroxidase) in a medium containing 50 mM Tris-HCl pH 7.5, added 150 mM NaCl and 5% skimmed milk, at a dilution of 1:3000. They were then washed three times in TBS buffer. The membranes were revealed according to the method described for peroxidase in Sambrook, et al.,(1989).

CHAPTER 4

RESULTS AND DISCUSSION

4.1 Respiration and ethylene profiles during banana ripening

Respiration (CO_2) and ethylene measurements were made in two independent experiments, the results of which are shown in Figures 6 and 7. The third experiment confirmed the results obtained in the second, where it can be seen that there were no differences between the profiles of the samples treated with GA_3 and the control, in terms of either respiration or ethylene. These results were different to those obtained by Vendrell (1970), who observed a delay in the respiratory peak when he infiltrated GA_3 (0.1 mM) into whole bananas, but the synthesis of ethylene observed by the author was not affected even at high gibberellin concentrations. However, the infiltration methodologies may explain the difference found. Vendrell (1970) used vacuum infiltration of slices and whole bananas (66 cm Hg/minute). The increase in ethylene production in the first 24 hours after infiltration shows that vacuum infiltration caused much greater stress on the tissue than the immersion system used in this work. The results shown in figure 7 make it clear that there was no significant increase in ethylene as a result of cutting and infiltration, which is in favour of the immersion infiltration method.

In other model systems, such as cell suspension in pear fruit, for example, a 33% reduction in the basal ethylene level in the culture medium was observed in those treated with GA_3 (1mM) after 3 hours of incubation at 25°C (Bem-Arie & Ferguson, 1991). This result provides some evidence that the endogenous gibberellin itself could be the signalling agent in delaying fruit ripening and not the effect caused by ethylene inhibition or peak respiration.

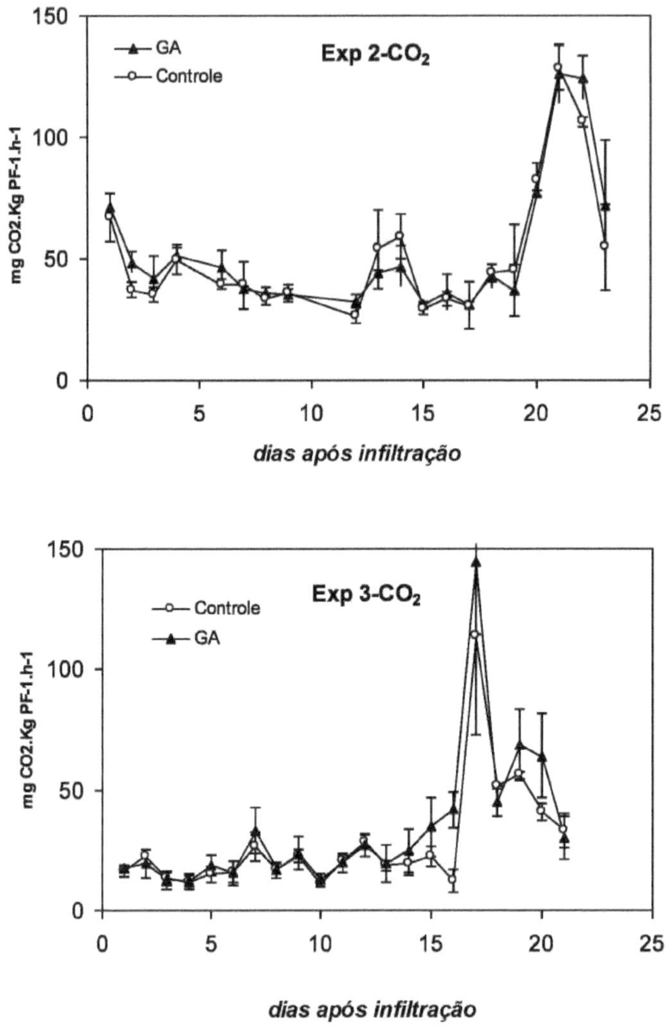

Figura 6: Respiration curves (mg CO_2.Kg PF-1.h-1) in banana slices treated with GA_3(0.1 mM+120 mM mannitol) compared to the control (120 mM mannitol)

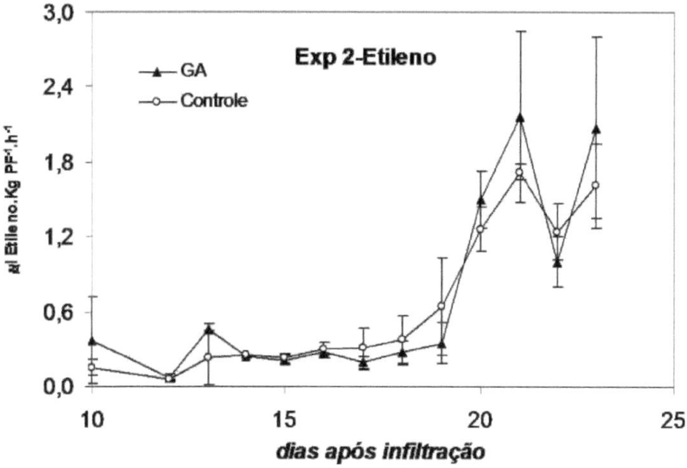

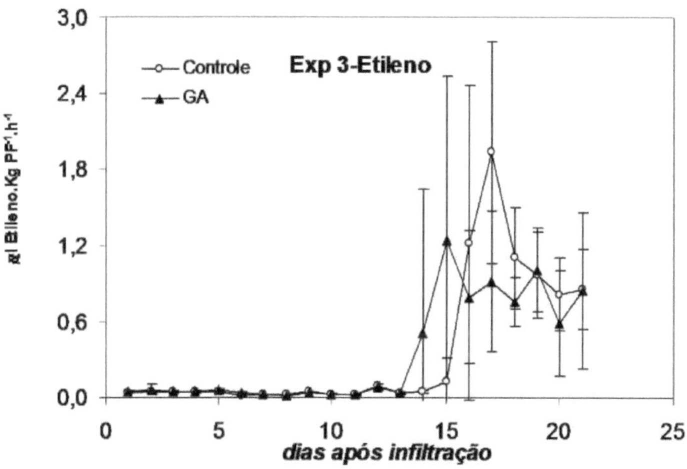

Figura 7: Ethylene profiles (Kg.PF^{-1}.h^{-1}) in banana slices treated with GA3 (0.1mM+120mM mannitol) and control (120 mM mannitol). Mean of triplicate extraction ± standard deviation

4.2 Starch and soluble sugar profiles during banana ripening

The delay in ripening was visible in the absence of the aroma of ripe bananas and the colour of the peel, which remained green for longer in the treated samples compared to the control. The delay in chlorophyll degradation caused by treatment with gibberellic acid has also been observed in citrus fruits (Biale, 1978; Abdel-Gawad & Romani, 1974), mangoes (Khader, 1992), strawberries (Martínez and coL, 1996) and parsley (Lers et aL, 1998).

The results shown in Figures 8 and 9 indicate that the samples treated with gibberellin A_3 had a delay of at least three days in the start of starch degradation and the accumulation of soluble sugars and, consequently, in the banana ripening process. However, once started, starch degradation/sugar accumulation proceeded normally as in the control slices. This is an unprecedented result, as there are no reports in the literature linking gibberellic acid and starch degradation.

Khader, (1991) demonstrated that high concentrations of gibberellic acid decrease the content of total soluble solids in mango. However, in fruit such as blueberries, where gibberellic acid is responsible for inducing parternocarpy, the accumulation of hexoses was not influenced and fruit development was delayed when compared to pollinated fruit (Medrano et aL, 1997). In nectarines, ripening was delayed with the application of gibberellic acid, in terms of skin colour and texture maintenance, but the soluble solids content was maintained, and the fruit therefore had better post-harvest quality, also increasing storage life by avoiding physiological and/or pathological disorders (Zilkah et aL, 1997). In mandarins, the soluble solids content was also not affected, and the quality of the product was also improved (Marur et aL, 1999). Unlike some of the data found in the literature, it was observed in this study that gibberellic acid delayed the accumulation of sugars, although at the ripe stage the fruit reached normal levels. The delay in starch degradation and sucrose accumulation observed in the treated samples suggests that the activity of the enzymes that degrade starch (amylases and phosphorylases) and/or the enzymes that synthesise sucrose (SPS and SS) may have been affected by gibberellic acid. This is why the activities of these enzymes were measured.

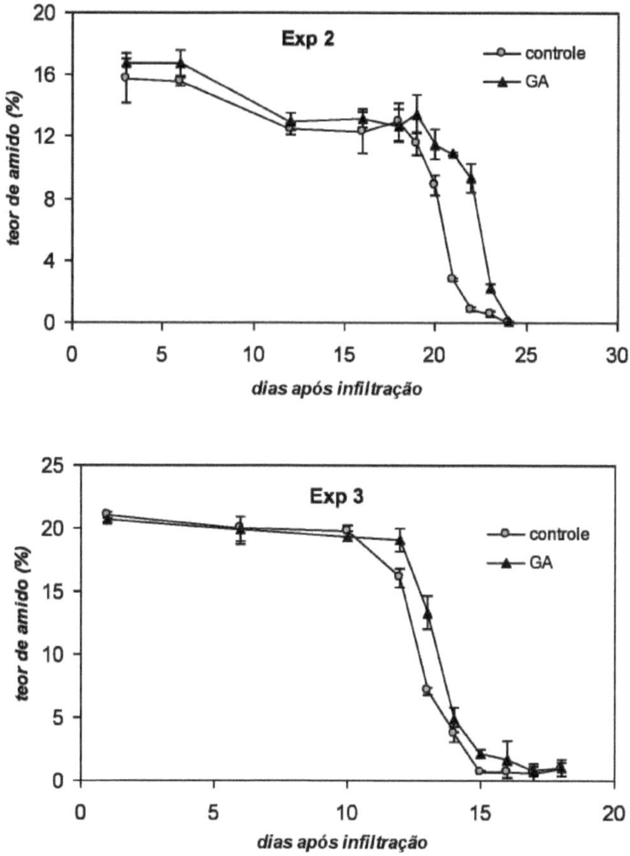

Figura 8: Starch profile in banana slices treated with gibberellin A_3 (0.1 mM GA_3 + 120 mM mannitol) and control (120 mM mannitol). Mean of triplicate extraction ±standard deviation.

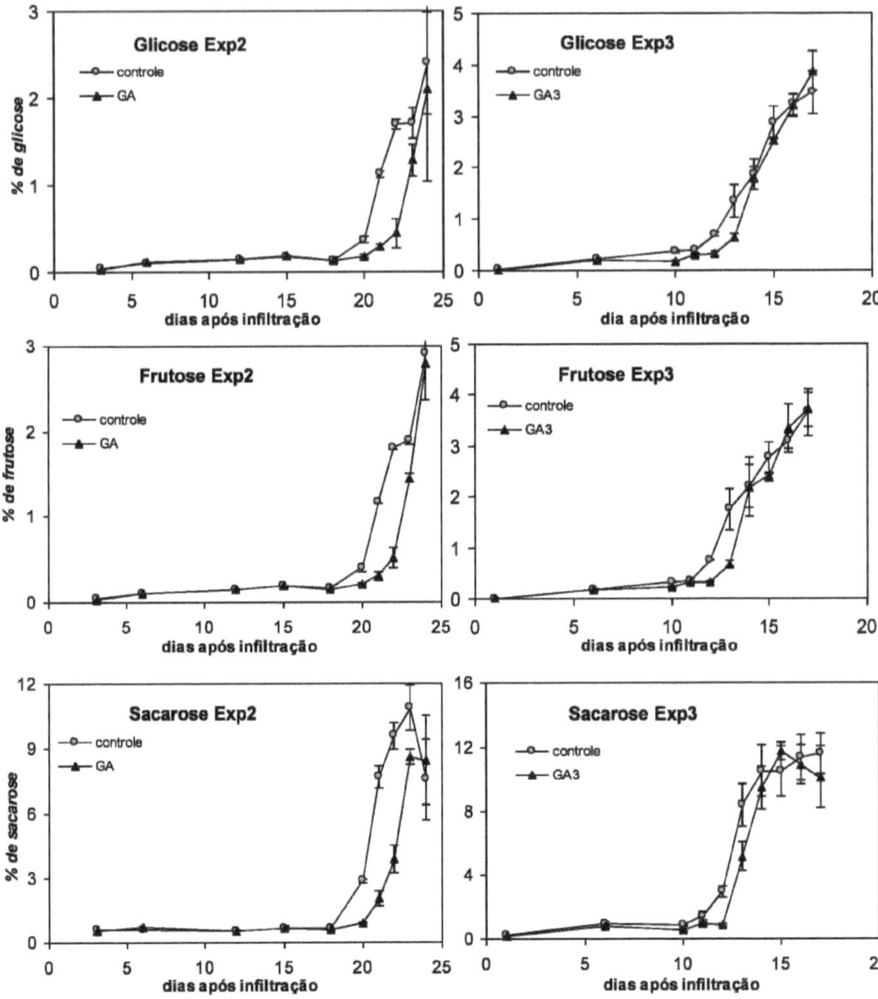

Figure 9: Glucose, fructose and sucrose content in control and treated banana slices, measured by HPLC. Average of triplicate extraction ± standard deviation.

4.3 Total hydrolytic activity

It is not yet known whether the enzymes that initiate the degradation of banana starch are phosphorylases or amylases. According to the literature, a-amylases, 0-amylases, phosphorylases, α-1,4 glucosidase and desramifying enzymes, such as isoamylases, take part in starch degradation (Arêas & Lajolo, 1981; Garcia & Lajolo, 1988). However, it is believed that only α-amylases are able to initiate the degradation of intact starch granules (Halmer, 1985), and the other enzymes complement their activity (Cardemil & Varner, 1984).

It is also known that all these enzymes are present and active in bananas during ripening (Garcia & Lajolo, 1988).

As can be seen in figure 10, no differences were observed between the total hydrolytic activity of the control and treated samples. The activity profile of the control samples was that normally found in whole bananas, with a small peak in activity on the 10th and 17th days and a slight drop in activity just before the start of starch degradation. The hormone seems to have had little effect on the normal behaviour of the enzymes considered to be hydrolytic. It is also not known whether the basal levels of activity of these enzymes would be sufficient to degrade the starch, or whether an increase in activity/expression would be necessary for degradation to occur.

The literature shows that amylase activity in avocado fruit was higher in young fruit than in ripe fruit. After harvest, an increase in amylase activity was observed near the climacteric peak, with a parallel reduction in the starch content of the flesh, suggesting that this profile is directly correlated to ripening processes such as increased respiration, increased ethylene content and increased softening (Pesis et al., 1978). The hydrolysis of starch during fruit ripening by hydrolytic enzymes has also been studied in pears (Pech & Latchke, 1972); mangoes (Mattoo & Modi, 1969) avocados (Pesis et al., 1978) tomatoes (Davis & Cocking, 1967), papayas (Pal & Selvaraj, 1987).

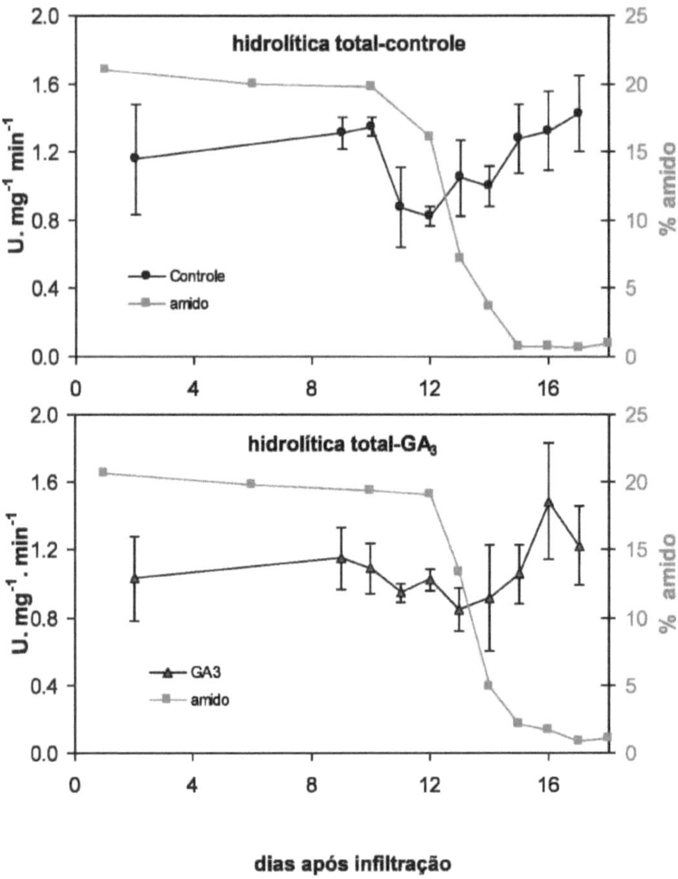

Figure 10: Total hydrolytic activity profiles (total amylases: α- and β-amylases, glycosidases and debranching) in banana slices from experiment 3°. Mean of triplicate extraction ±standard deviation

4.4 . α-amylase activity

The data illustrated in figure 11 show that there is no significant increase in α-amylase activity associated with the degradation of starch and the accumulation of soluble sugars, both in the control and treated samples. This result differs from that found in tomatoes (Davies & Cocking, 1965) and kiwi fruit (Wergrzyn & Macrae, 1995) where an increase in the activity of this enzyme was observed, associated with the period of rapid starch hydrolysis.

Despite having very similar profiles, the samples treated with gibberellin A3 had an initial activity around 2.5 times lower than the control samples (1U), and the maximum activity was 1.5 U. The control samples always had at least twice as much α-amylase activity as the samples treated with gibberellic acid. This difference could, in part, explain the delay in the onset of starch degradation in the slices treated with gibberellic acid and which, due to differences in activity methods, is not reflected in the total hydrolytic activity (amylases, glycosidases and desramifiers) as shown in the previous paragraph.

Khader (1992) observed similar results in mango, which shows that fruit can behave very differently in relation to a-amylase when compared to cereals, for example, where the literature amply demonstrates that gibberellic acid causes an increase in the gene expression of α-amylase isoforms (Davies, 1997; Perata et al., 1997; Loreti, 1998; Rogers & Rogers, 1999). It has been observed in rice cells that, in addition to gibberellic acid, sugar concentration can also regulate protein synthesis and the post-translational induction of a-amylase, through the control of calcium channels (Mitsui et al., 1999).

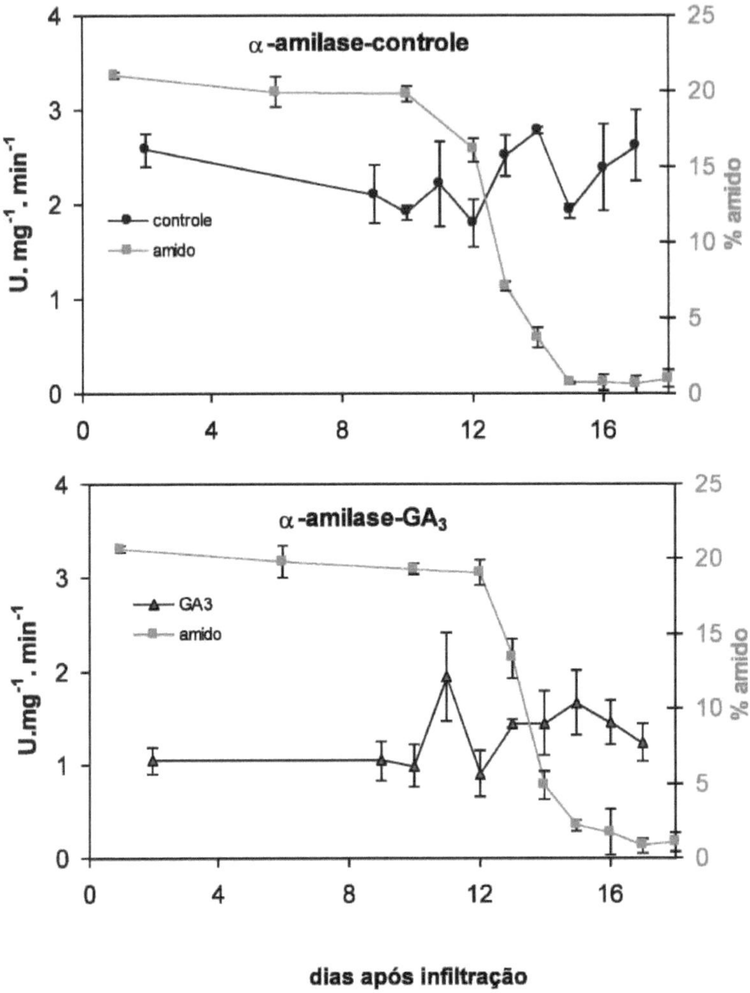

Figure 1: Profile of specific α-amylase activity in banana slices, relative to the 3rd experiment. Average of triplicate extractions ± standard deviation.

4.5 β-amylase activity profile

The data in figure 12 shows that the treatment with gibberellin A_3 caused a significant delay in the increase in β-amylase activity compared to the control slices, from the 18th day onwards. This delay of at least three days coincides with the delay in starch degradation in the treated samples. Although β-amylase is not fundamental in starch hydrolysis, it is significant that it was affected by treatment with gibberellic acid together with a-amylase. The results found here are different to those found for cereals such as rice, where it has been observed that the application of gibberellic acid does not increase the activity of β-amylase in germinating grains, but can reverse the activity of the enzyme when it is inhibited by abscisic acid (Wang et al., 1996). The data observed in figure 12 is similar to that found by Purgatto et al. (2001), who infiltrated IAA into banana slices.

β-amylase is present and appears to play an important role in the mobilisation of starch during the germination process and in the production of malt in some cereals, such as sorghum (Uriyo & Eigel, 1999) and barley (Kihara et al., 2000). However, studies on *Vigna unguiculata* showed that this enzyme was not present during the germination phase, and that α-amylase was the enzyme responsible for this process (Uriyo, 2001). In figs, which contain a low amount of starch, β-amylase, along with α-amylase, also appears to play an important role in carbohydrate metabolism, with a peak in activity being found 10 days before ripening when the fruit was attached to the plant (Yoshioka, 1995).

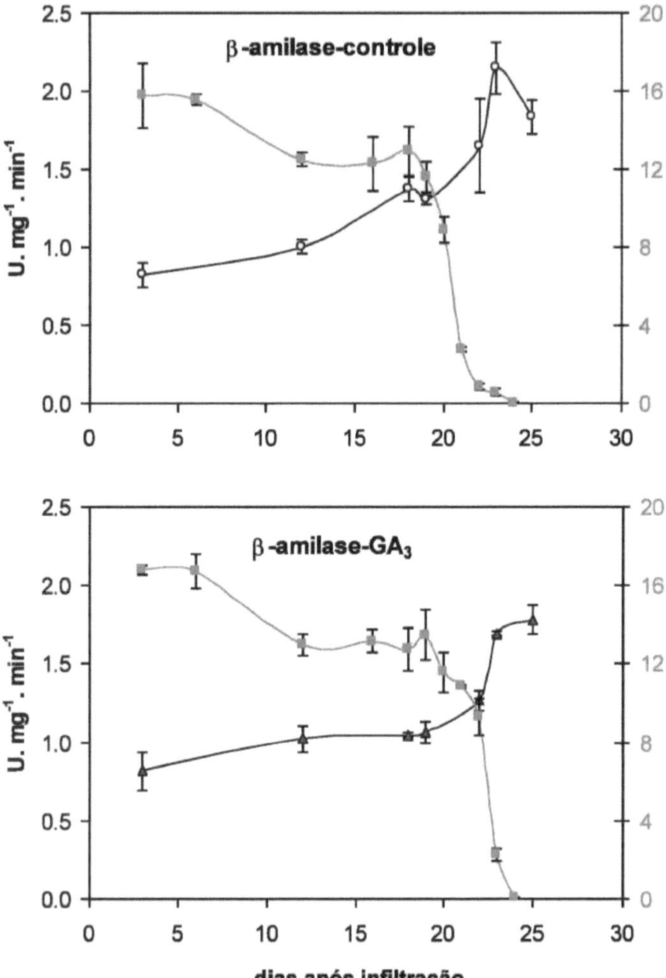

Figure 2: Profile of specific β-amylase activity in banana slices, relative to experiment 2(o). Mean of triplicate extraction ±standard deviation.

4.6 Phosphorylase Activity Profiles

The enzyme starch phosphorylase (EC 2.4.1.1., α-1,4-D-glucan: ortho-phosphate a-glucosyl transferase) catalyses the reversible conversion of starch into G-1P in the presence of orthophosphate, hydrolysing the terminal α-1,4 bond. It is widely distributed in plants as a single or multiple form (Nighojkar and Kumar, 1997). High levels of phosphorylase activity were found in kiwi fruit, indicating significant activity of the enzyme during fruit development and ripening. A relatively constant level of amylolytic and phosphorolytic activity was observed during the period when the starch content increased, decreasing significantly thereafter (Wergrzyn and Macrae, 1995).

It has already been observed that the activity of phosphorylases in bananas increases slightly before the onset of starch hydrolysis, decreases during the climacteric period and increases again during senescence (Arêas and Lajolo, 1981, Mota, 2001). Despite being discrete variations, the profiles are reliable as they have been repeated several times. Other studies on bananas by Goswami and Borthakur (1996) also showed an increase in phosphorolytic activity in the direction of starch synthesis.

The data obtained in this study, shown in Figures 13.a and 13.b, showed that the phosphorylase activity profiles of the control and treated samples did not differ and were similar to those normally observed by Arêas and Lajolo (1981) and Mota (2001). The data obtained from the third experiment confirmed the results of the second. These results are reinforced by western blotting against anti-banana phosphorylase antibody (Mota, 2001) (figures 14 and 15) where it can be seen that there really do not seem to be any differences between control and treated samples in terms of protein expression.

The results show that amylases seem to be affected by gibberellic acid and phosphorylases do not, which reinforces the idea that it is amylases and not phosphorylases that initiate starch degradation in bananas.

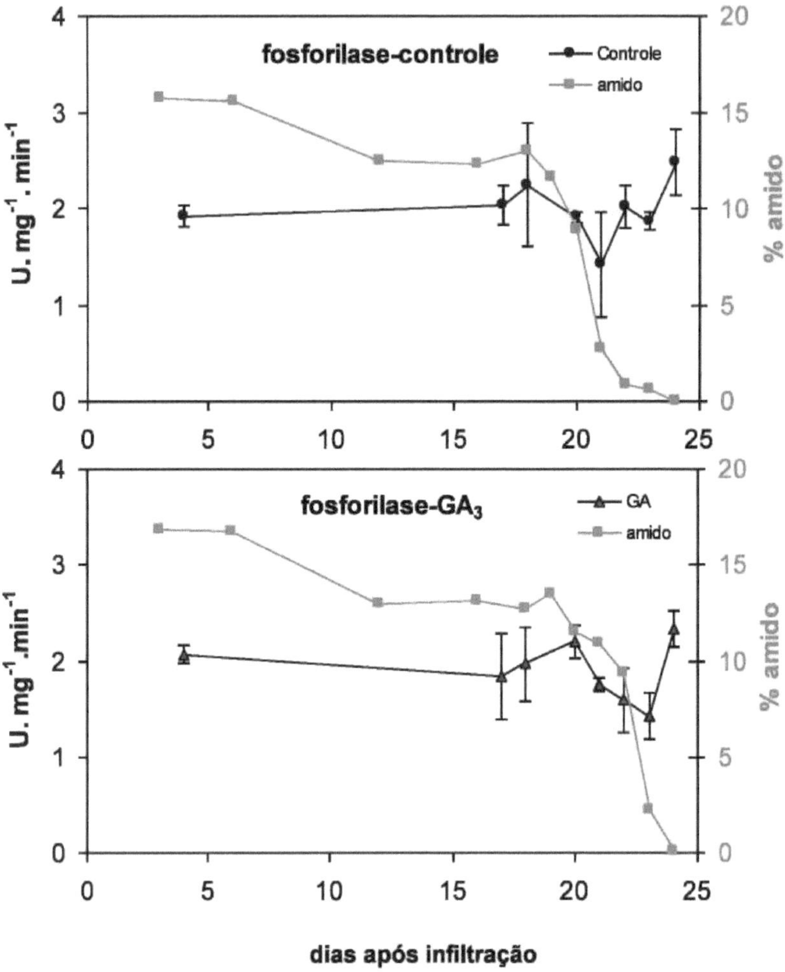

Fig.13.a: Phosphorolytic activity in banana slices, relative to experiment 2º, in relation to the starch curve. Average of three extractions ± standard deviation

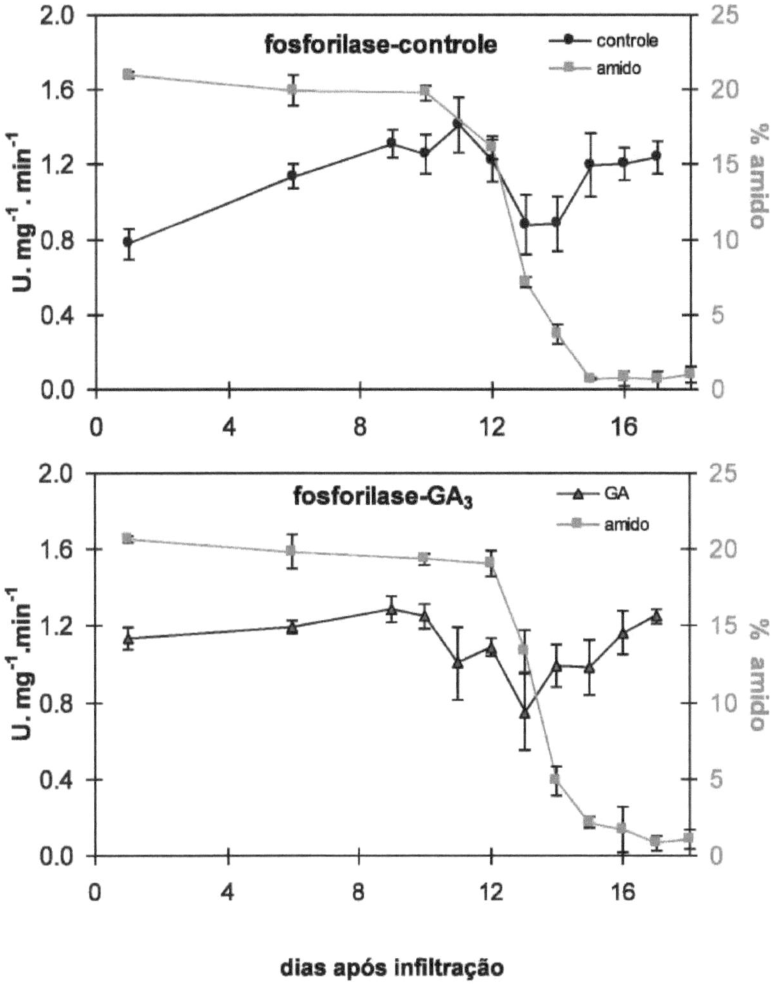

Fig.13.b: Phosphorolytic activity in banana slices, relative to experiment 3°, in relation to the starch curve. Average of three extractions±standard deviation.

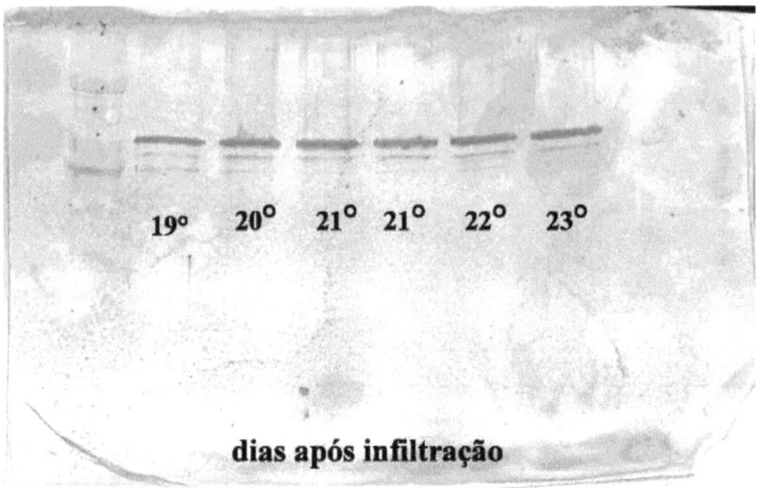

Figura 14: Western-blotting revealed with anti-phosphorylase antiserum in control banana slices (120 mM mannitol) during ripening.

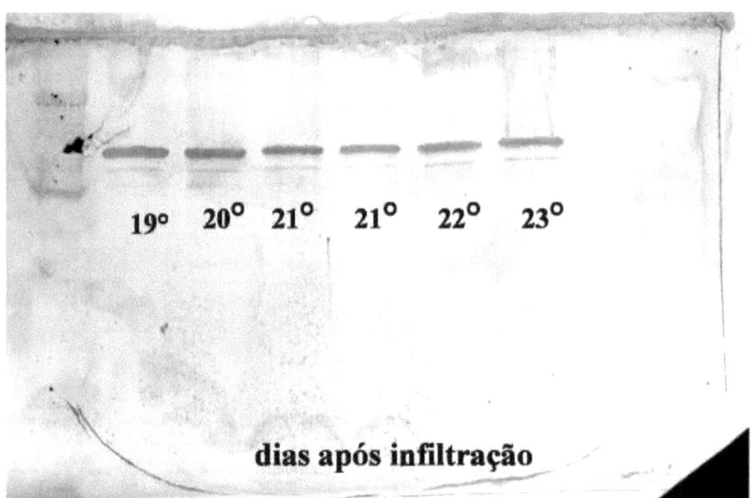

Figura 15: Western-blotting revealed with anti-phosphorylase antiserum in banana slices treated (0.1 mM GA3 + 120 mM mannitol) during ripening.

4.7 Sucrose phosphate synthase activity profile

The SPS activity profile in the control samples was similar to that found by Cordenunsi (1989) and Nascimento et al (1997), where there is a correlation between the increase in enzyme activity and the increase in sucrose content. However, the activity of

SPS in the samples treated with GA_3 seems to have been affected by the hormone, since there was a decrease in activity, around two times lower on the 20th and 21st day after infiltration. Western blotting (figures 17 and 18) showed a lower amount of protein on the 21st day in the treated samples, coinciding with the lower activity observed during the period analysed. However, based on the total protein gels (Figures 19 and 20), it could not be ruled out that a lower amount of protein had been applied at this point.

Nascimento et al (1997) proved that there is an increase in SPS expression throughout ripening, which begins before harvest and extends throughout the banana ripening process, concomitantly with the increase in activity and the accumulation of sucrose. From the results obtained by Western blot (Figures 19 and 20) against the SPS antibodies (Nascimento et al., 1997), it was not possible to verify an increase in SPS expression, probably because the period analysed here was shorter and because of problems with the methodology.

There are not many records relating gibberellic acid levels and SPS activity, but Cheikh et al. (1992) identified a possible signal in the regulation of SPS activity by gibberellic acid, which would be acting indirectly by modifying the levels of glucose-6-P and Pi, considered allosteric regulators of the enzyme.

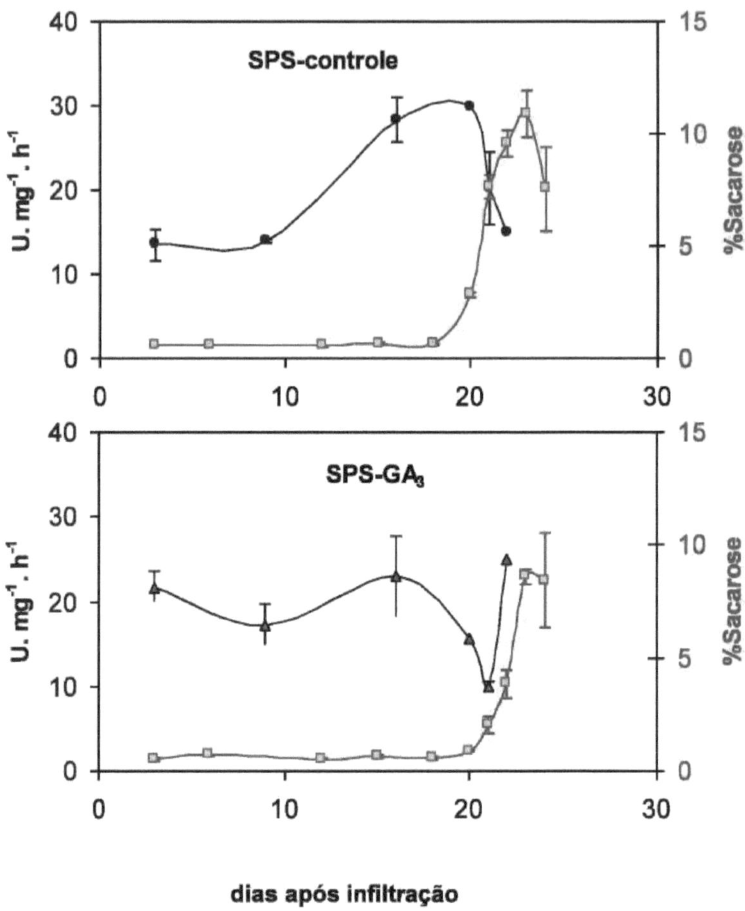

Figura 16: Sucrose phosphate synthase activity, relative to experiment 2º, in bananas treated with gibberellic acid compared to control slices and the sucrose profile. Mean of triplicate extraction ± standard deviation

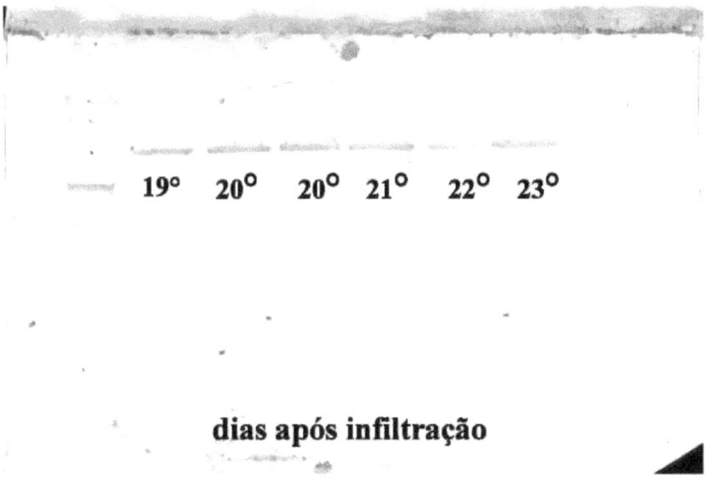

Figura 17: Western-blotting revealed with anti-SPS serum on control banana slices (120 mM mannitol) during ripening

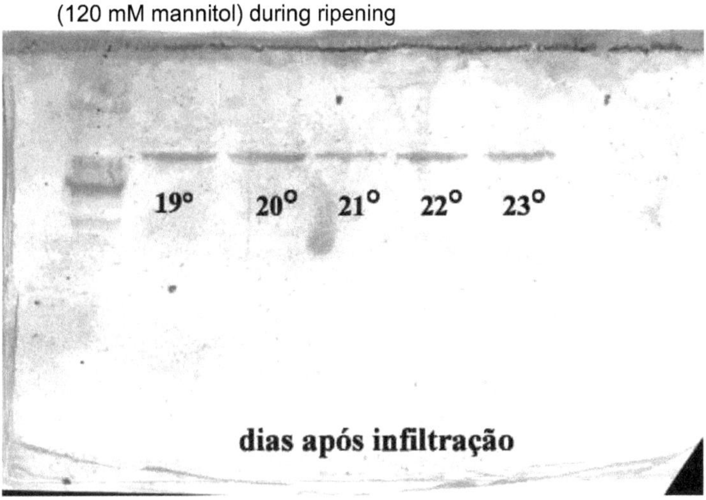

Photo 18: Western-blotting revealed with anti-SPS serum on banana slices treated (0.1 mM GA3+ 120 mM mannitol) during ripening

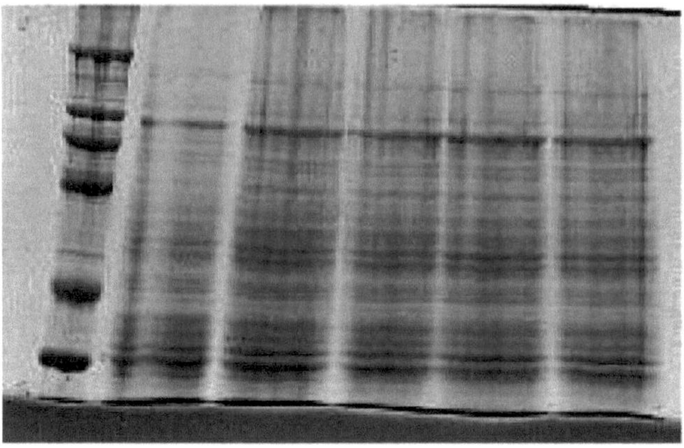

Figura 19: Dissociating gel electrophoresis (SDS-PAGE) on control banana slices (120 mM mannitol)

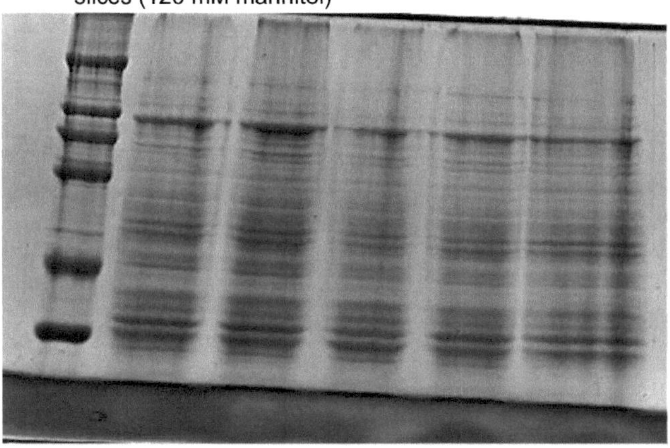

Figura 20: Dissociating gel electrophoresis (SDS-PAGE) on banana slices treated (0.1 mM gibberellic acid +120

4.8 SS enzyme activity profile

Sucrose synthase (SS) acts reversibly in the synthesis and hydrolysis of sucrose (Geigenberger & Stitt, 1993). In bananas, it is highly active during fruit formation and practically disappears during ripening, when sucrose is synthesised and accumulated. It therefore appears to be more closely linked to the hydrolysis of sucrose during the fruit formation period to form cell walls and starch (Cordenunsi and Lajolo, 1995). In this study, SS activity was measured to see if gibberellic acid could halt this loss of activity. The results showed that there is no such reversal since the activity profiles (figure 15) were very similar in slices treated with gibberellin A_3 and the control, despite the fact that, like SPS, there was a relatively higher initial activity in the case of the treated slices. Also like SPS, this activity equalises around the 10th day after infiltration and increases on the 20th day. These results can be compared to those obtained with blueberries, where the authors observed an increase in SS activity in the ripening phase (at 85 days after anthesis) (Medrano et al., 1997).

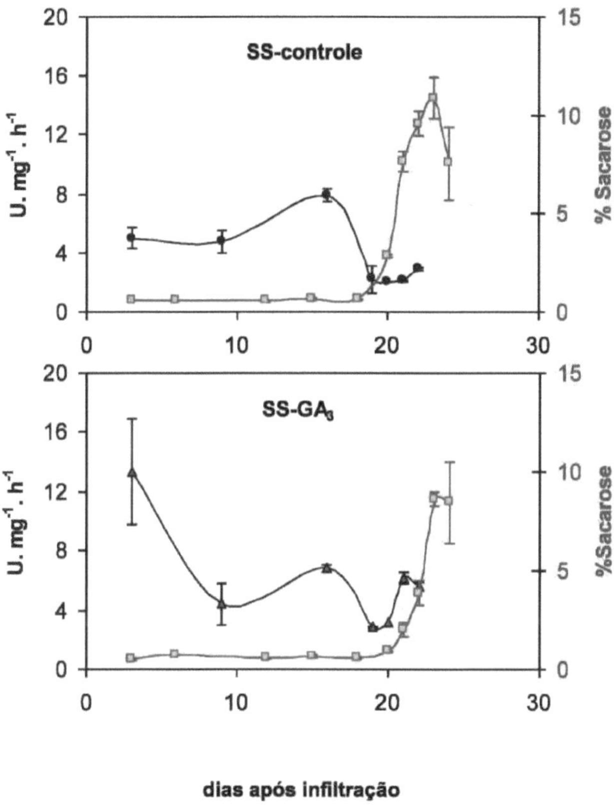

Figura 21: Sucrose synthase activity in bananas treated with gibberellic acid compared to control slices and the sucrose profile. Mean of triplicate extraction ± standard deviation

CHAPTER 5

FINAL CONSIDERATIONS

The slice study model proved to be suitable, as the respiratory activity, ethylene production and carbohydrate profiles: starch and sugars, were similar to the profiles already obtained with the whole banana. Through the analyses of respiration and ethylene, it was observed that infiltration in the absence of vacuum does not appear to have caused significant stress to the tissue.

It was observed that gibberellic acid did not affect respiratory climacteric and did not interfere with ethylene production.

Three independent experiments have shown that gibberellic acid delays ripening, as seen by the two to three day delay in starch degradation and the accumulation of soluble sugars.

The activity of some of the enzymes that degrade starch and synthesise sucrose seems to have been affected by gibberellic acid. The activity of a- amylase was around twice as low in slices treated with GA_3 compared to the control. This decrease could partly explain the delay in starch degradation. Both the activity and expression of phosphorylase do not seem to have been significantly affected by infiltration with gibberellic acid. SPS activity was about two times lower at the 20th and 21st dpi in treated slices, but there was no decrease in protein expression of this enzyme.

CHAPTER 6

CONCLUSIONS

- The ripening of banana slices treated with gibberellic acid was delayed by at least three days. This was mainly due to the delay in starch degradation and the accumulation of soluble sugars in the pulp;
- Amylases seem to have been affected by treatment with gibberellic acid, but phosphorylases have not;
- It is not possible to say that sucrose-phosphate synthase and sucrose synthase were affected by the treatment.

BIBLIOGRAPHICAL REFERENCES

ABELES, F. B; MORGAN, P. W.; SALTVEIT, M. E. **Ethylene in plant biology.** Ed. Academic Press, London. 2ª ed., 414p., 1992.

ADAMS, D.O., YANG, S.F. Ethylene biosynthesis: Identification of 1- aminocyclopropane-1-carboxylic acid as an intermediate in the conversion of methionine to ethylene. **Proc. Natl. Acad. Sei.,** USA, v.76, p. 170-4, 1979.

ARÊAS, J. A. G., LAJOLO, F. M. Starch transformation during banana ripening: I- The phosphorylase and phosphatase behovior in Musa acuminata. **J. Food Biochem.** Wesport, v.5, p. 19-37. 1981.

AP REES, T. Integration of pathways of synthesis and degradation of hexose phosphates. In: STUMPF, P. K.; CONN, E. E. The *biochemistry of Plants.* New York: Academic Press, 1980, v.3, p. 1-42.

BALL, K. L; GREEN, J.H.; AP REES, T. Glycolysis at the climacteric in bananas. **Eur. J. Biochem.,** v. 177, p.637-641, 1991.

BASSINELLO, P. Z.; FIORAVANTE, A. P.; NASCIMENTO, J.R.O. Distribution of sucrose-phosphate synthase and sucrose synthase in bananas during ripening. **Ciênc. Tecnol. Aliment.,** Campinas, v. 19, n. 1, p. 102-106, 1999.

BEN-ARIE, R.; ROISMAN, YALA, ZUTHI, Y. Gibberellic acid reduces sensitivity of persimmon fruits to ethylene, p. 165-171. In: CLIJSTERS, H.; DE PROFT, M.; MARCELLE, R.; VAN POUCK, M. *Biochemical and Physiological Aspects of Ethylene Production in Lower and Higher Plants.* Boston: Kluwer Academic Publisher, 1989, p. 165-171.

BEN-ARIE, R. & FERGUSON, B. lan. Ethylene production by growing and senescing pear fruit cell suspensions in response to gibberellin. **Plant Physiol.,** Rockville, v.95, p. 943-947, 1991.

BERGMEYER, H. U., ed **Methods of enzymatic analysis,** Ed. Academic Press, New York, 2ed" v.3: 1212-1215. 1984.

BIONDI.S., DIAZ, T" IGLESIAS, I., GAMBERINI, G" BAGNI, N. Polyamines and ethylene in relation to adventitious root formation in *Prunus avium* shoot cultures. **Physiol. Plant,** Kobenhavn, v.78, p.474-83, 1990.

BRADFORD, M. M. A rapid and sensitive method for the quantification of microgram quantities of protein utilising the principie of protein-dye binding. **Analytical Biochemistry,** v. 72, p. 248-54, 1976.

BRUNEU, J. M.; WORRELL, A. C" CAMBON, B.; LANDO.D.; VOELKER, T. A. Sucrose phosphate synthase, a key enzyme for sucrose biosynthesis in plants. **Plant Phisiology**, Kobenhavn, v. 96, p. 473-478, 1991.

BUCHNER, P. BORISJUK, L., WOBUS, U. Glucan phosphorylases in *Vicia faba* L.: cloning, structural analysis and expression patterns of cytosolic and plastidic forms in relation to starch. **Planta,** Berlin, v. 199, p. 64-73, 1996.

CARAWAY, W. T. A stable starch substrate for the determination of amylase in serum and other body fluids. **American Journal of Clinical Pathology,** v.32 (1), p.97-99, 1959.

CARDEMIL, L.; VARNER, J.E. Starch degradation metabolism towards sucrose synthesis in germinating *Araucaria araucana* seeds. **Plant Physiol.,** Rockville, v. 76, p. 1047-1054, 1984.

CORDENUNSI, B. R.; LAJOLO, F. M. Starch breakdow during banana ripening: sucrose synthase and sucrose phosphate synthase. **J. Agric. Food Chem.** Washington, v. 43 (2), p. 347-351, 1995.

CORDENUNSI, B. R. **Sucrose synthesis during banana ripening. Involvement of sucrose synthase and sucrose phosphate synthase.** São Paulo, 1989. 87p. (Doctoral Thesis - Faculty of Pharmaceutical Sciences - University of São Paulo).

CHITARRA, A. B. **Contribution to the study of the physiology and post-harvest biochemistry of the "Marmelo" banana.** São Paulo, 1979. 110p. (Doctoral Thesis - Faculty of Pharmaceutical Sciences - USP)

DAVIES, J. W.; COCKING, E. C. Changes in carboydrate, proteins and nucleic acids during cellular developments in tomato fruit locule tissue. **Planta,** Berlin, v.67, p. 242-253, 1965.

DAVIES, J. W.; COCKING, E. C. Protein synthesis in tomato fruit locule tissue; the sites of synthesis and the pathaway of carbon into protein. **Planta,** Berlin, v.76, p. 285-305, 1967.

DAVIES, P. J. *PLANT HORMONES: Physiology, biochemistry and molecular biology.* Boston: Kluewer Academic Publishers, 1995.

DAVIES, C.; BOSS, P.K.; ROBINSON, S.P. Treatment of grape berries, a nonclimacteric fruit with a synthetic auxin, retards ripening and alters the expression of developmentally regulated genes. **Plant Physiol.,** Kobenhavn, v. 115, p. 1155-1161, 1997.

DEY, P. M.; DIXON, R. A. *Biochemistry of storage carbohydrates in green plants,* London : Academic Press, 1985, 378p.

DOEHLET, D. C. HUBER, S. C. Spinach leaf sucrose phosphate synthase. Activation by glucose 6-phosphate and interaction with inorganic phosphate. **FEBS Lett.,** Amsterdam, v. 153, n. 2, p. 293-297, 1983 (a).

DOEHLET, D. C. HUBER, S. C. Regulation of spinach leaf sucrose phosphate synthase by glucoseõ-phosphate, inorganic phosphate and pH. **Plant Physiology,** Bethesda, v. 73, p. 989-994, 1983 (b).

DUNN, G. A model for starch breakdown in higher plants. **Phytochemistry**, Kiel, v. 13, p. 1341-1346, 1974.

EVEN-CHEN, Z" MATOO, A.K., GOREN, R. Inhibition of ethylene biosynthesis by aminoethoxyvynilglycine and by polyamines shunts label from 3,4-[^{14}C] methionine into spermidine in aged peei discs. **Plant Physiology,** Bethesda, v.69, p.385-88, 1982.

FOSKET, D. E. *Plant growth and development: a molecular approach.* San Diego: Academic Press, 1994, 580p.

GALLANT, D.J.; BOUCHET, B.; BALDWIN, P. M. Microscopy of starch: evidence of a new levei of granule organisation. **Carbohydrate Polymers** v, 32, p. 177-191, 1997.

GARCIA, E. **Transformations of the amyloplast and participation of amylolytic enzymes in banana ripening.** Dissertation (Master's Degree), Faculty of Pharmaceutical Sciences, University of São Paulo, São Paulo, 72 p, 1983.

GARCIA, E., and LAJOLO, F.M. Starch transformation during banana ripening: the amylase and glucosidase behaviour. **Journal of Food Science,** v. 53 (4), p. 11811186, 1988.

GARCIA-MARTÍNEZ. L. J & HEDDEN, P. Gibberllins and fruit development. In: TOMÁS-BARBERÁN, F. A.; ROBINS, R. J. *Phytochemistry of Fruit and Vegetables.* USA: Oxford University, p. 263-285, 1997.

GEINGENBERGER, P. & STTIT, M. Sucrose synthase catalyses a readily reversible reaction in vivo in developing potato tubers and other plant tissues. **Planta,** Berlin, v, 189, p. 329-339, 1993.

GILLASPY, G.; HILLA, B.D.; GRUÍSSEM, W. Fruits: a developmental perspective. **The Plant Cell,** California, v.5, p. 1439-1451, 1993.

GUBLER, F.; KALLA, R.; ROBERTS, K. J; JACOBSEN, V. J. Gibberellin-regulated expression of a myb gene in barley aleurone cells: evidence for Myb transactivation of a high-pl α-amylase gene promoter. **The Plant Cell,** California, v. 7, p. 1879-1891, 1995.

HAGENIMANA, V.; VÈZINA, L. SIMARD, R.E. Distribution of amylases within sweet potato (Ipomes batatas L.) root tissue. **J. Agric. Food Chem.,** v. 40, p. 1777-1783, 1992.

HALMER, P. The mobilisation of storage carbohydrates in germinated seeds. **Physiol. Veg.,** v. 23, p. 107-125, 1985.

HAMILTON, A. J.; LYCETT, G. W; GRIERSON, D. Antisense gene inhibits synthesis of the hormone ethylene in transgenic plants. **Nature,** London, v. 346, p. 284-287, 1990.

HARBERD, P. N.; KING, E. K.; CAROL, P.; COWLING, J. R.; PENG, J.; RICHARDS, E. D. Gibberellin: inhibitor of na inhibitor of...?. **BioEssay,** v. 20, p. 1001-1008, 1998.

HESSE, H.; SONNEWALD, U.; WILLMITZER, L. Cloning and expression analysis of sucrose phosphate synthase from sugar beet (Beta *vulgaris* L). **Mol. Gen. Genet.,** v. 247, p. 515-520, 1995.

KAYISU, K.; HOOD, L. F" VANSOEST, P. J. Characterisation of starch and fibre of bananafruit. J. **Food. Sei.,** Chicago, v. 46, p. 1875-1890, 1981.

KRUCKEBERG, A. L; NEUHAUS, H. E.; FEIL, R.; GOTTLIEB, L. D" STITT, M. Decreased-activity mutants of phosphoglucose isomerase in the cytosol and chloroplast of Clarkia xantiana- Impact on mass action rations and fluxes to sucrose and starch, and estimation of flux control coefficients and elasticity coefficients. **Biochem. J.** London, v. 261, p. 457-467, 1989.

KHADER, S. E. S. A.; SINGH, B. P" KHAN, S. A. Effect of GA3 as a post-harvest treatment of mango fruit on ripening, amylase and peroxidase activity and qualit during storange. **Science Hort.,** v. 36, p. 261-266. 1988.

KHADER, S. E. S. A. Effect of preharvest application of GA3 on postharvest behaviour of mango fruits. **Science Hort.,** v. 47, p. 317-321, 1991

KHADER, S. E. S. A. Effect of gibberellic-acid and vapour gard on ripening, amylase and peroxidase-activities and quality of mango fruits during storage.**Journal of Horticultural Science, v.** 67(6): p. 855-860, 1992.

KONISHI, Y.; HARADA, M.; D' INNOCENZO, M.; LAJOLO, M. F. Purification and characterisation of soluble and cell wall-bound acid a-glucosidases of ripe yellow banana pulp. **J. Appl. Glycosci.,** v. 48 (1): p. 19-25, 2001.

KUMAR, A. & SANWAL, G. G. Purification and physicochemical of starch phosphorylase from young banana leaves. **Biochemistry,** v. 21, p. 4152-4159, 1982.

HUBBARD, N. L., PHARR, D. M., HUBER, S. C. Role of sucrose phosphate in sucrose biosynthesis in ripening bananas and its relatioship to the respiratory climacteric. **Plant Physiology,** Bethesda, v. 94, p. 201-208. 1990.

HUBBARD, N. L., PHARR, D. M., HUBER, S. C. Sucrose phosphate synthase and other metabolising enzymes in fruits of various species. **Physiol. Plant.** Copenhagen, v. 82, p. 191-196, 1991.

LAEMMLI, U. K. Cleavage of structural proteins during assembly of the bacteriophage T4. **Nature,** London, v. 94, p. 201-208, 1970.

LEHAV, E. Value of K-Ca+Mg ration for determination of nutritional-status of banana sucker. **Fruit,** v. 33 , n. 1, p.3-6, 1978.

LELOIR, L. F.; CARDINI, C.E. The biosynthesis of sucrose phosphate. **J. Biol. Chem.,** Baltimore, v. 214, p. 157-165, 1955

LERS, A.; JIANG, W.B.; LOMANIEC, E.; AHARONI, N. Gibberellic acid and CO2 additive effect in retarding postharvest senescence of parsley. **Journal of Food Science,** v. 63 (1): p. 66-68, 1998.

LORETI, E.; GUGLIELMINETTI, L.; YAMAGUCHI, J.; GONZALI, S. Effect of anoxia on gibberellic acid-induced protease and ?-amylase Processing in barley seeds. **J. of Plant Physiol.,** v. 152, p. 44-50, 1998.

LOWRY, H. O. & LOPEZ, A. J. The determination of inorganic phosphate in the presence of labile phosphate ethers. **J. Biol. Chem.,** Baltimore, v. 162, p. 421-428 , 1946.

LOWRY, O.H.; ROSENBROUGH, N. J.; RANDHALL, R. J. Protein measurement with the Folin phenol reagent. J. **Biol. Chem.,** Baltimore, v.193, p. 265-275, 1951.

MACRAE, E" QUICK, W. P" BENKER, C" STITT, M. Carbohydrate metabolism during postharvest ripening in kiwifruit. **Planta,** Berlin, v. 188, p. 314-323, 1992.

MATOO, A. K.; MODI, V. V. Ethylene and ripening of mangoes. **Plant Physiol,** Bethesda, v. 44, p. 308-310, 1969.

MARTÍNEZ, G. A.; CHAVES, A. R.; and ANÓN, M. C. Effect of Exogenous Application of Gibberellic Acid on Colour Chage and Phenylalanine Ammonia-lyase, Chlorophyllase, and Peroxidase Activities during Ripening of Strawberry Fruit. **Journal of Plant Growth Regulation,** v.15, p. 139-146, 1996.

MARUR, C.J.; STENZEL, N.M.C.; RAMPAZZO, E.F.; SCHOLZ, M.B.S. Gibberellic acid (GA3) and ripening of Ponkan and Montenegrina mandarin fruits. **Scie. Agr.** v. 56 (3) 1999.

McCLEARY, B. V.; CODD, R. Measurement of 0-amylase in cereal flours and commercialenzyme preparations. **J. Agric. Food Chem.** V. 9, p. 17-33, 1989.

McGLASSON, W.B.; WADE, N.L.; ADATO, I. Phytohormones and fruit ripening. **In: Phytohormones and Related Compounds - A Comprehensive Treatise.** Ed. Letham, Goodwin and Higgins, v. 2, p. 447-493. Elsevier/North Holland Biochemical Press.

McKEON, T.S., FERNANDEZ-MACULET, J.C., YANG, S.F. Biosynthesis and metabolism of ethylene. In: DAVIES, P.J. *PLANT HORMONES: physiology, biochemistry and molecular biology.* Dordrecht: Kluwer Academic Publish., p. 118204, 1995.

MENDICINO, J. Sucrose-phosphate synthesis in wheat germ and green leaves. **J. Biol. Chem. Baltimore,** v. 235 (12): p. 3347-3352, 1960.

MEDRANO.C. R. & DARNELL, L. R. Sucrose metabolism and fruit growth in parthenocarpic vs seeded blueberry */Vaccinium ashei)* fruits. **Physiol. Plant,** Bethesda, v. 99, p. 439-446, 1997

MITSUI, T.; LOBODA, T.; ITOH, A.; IKARASHI, T. Sugar-controlled Ca2+ uptake and alpha-amylase secretion in cultured cells of rice *{Oryza sativa* L.). **Plant and Cell Physiology,** v. 40 (8): p. 884-893, 1999.

MOTA, R. V; LAJOLO, F.M. & CORDNUNSI, B.R. Carbohydrate composition of some banana cultivars (Musa spp.) during ripening. **Ciênc. Tecnol. Aliment.,** Brazil, v. 17 (2): p. 94-99, 1997.

MONROE, J. D.; PREISS, J. Purification of a 0-amylase that accumulates in Arabidopsis thaliana mutants defective in starch metabolism. **Plant Physiology,** Bethesda, v.94, p. 1033-1039, 1990.

MOTA, R. V. **Starch-sucrose metabolism and determination of soluble sugars in some banana cultivars** *{Musa spp).* São Paulo, 1997. 75p. (Master's Thesis - Faculty of Pharmaceutical Sciences - University of São Paulo).

NASCIMENTO, J. R. O; CORDENUNSI, B.R.; LAJOLO, F.M. and ALCOCER, M.J.C. Banana sucrose-phosphate synthase gene expression during fruit ripening. **Planta,** Berlin, v. 203, p. 283-288, 1997.

NASCIMENTO, J. R. O; CORDENUNSI, B.R.; LAJOLO, F.M. Sucrose synthase activity and expression during development and ripening in banana. **J. of Plant Physiol.,** v. 156, p. 605-611, 2000.

NIGHOJKAR, K., KUMAR, A. Starch phosphorylase: biochemical, molecular and biotechnological aspects. **The Genetic Engineer and Biotechnologist. v. 17** (4): p. 189-202. 1997.

OELLER, P.A.; MIN-WONG, L.; TAYLOR, L.; PIKE, DA; THEOLOGIST, A. Reversible inhibition of tomato fruit senescense by antisense RNA. **Science,** v. 254, p. 437-439, 1991.

PAGE, A.E., WILLIANS-ASHMAN, H.G On the role of S-adenosyl-L-methinine in the biosynthesis of spermidine by rat protoplast. **J. Biol. Chem.,** v. 244, p.682-93, 1969.

PECH, J. C.; LATCHKE, A. Activities of enzymes involved in sugar metabolism in Passe-

Crassane pears during cold storage. J. **Sei. Food Agric.,** v. 23, p. 14991502, 1972.

PERATA, P.; MATSUKURA, C.; VERNIERI, P.; YAMAGUCHI, J. Sugar repression of a gibberellin-dependent signalling pathway in barley embryos. **American Society of Plant Physiologists,** v. 9, p. 2197-2208, 1997.

PERCHERON, F. Dosage colorimetric des fructose et des fructofuranosides par l'acide thiobarbiturique.C. **R. Acad. Scien**., Paris, v. 255, (19), p. 2521, 1962.

PERES, F. J.; MORALES, V. A basic peroxidase isoenzyme from de grape pedicel is induced by gibberellic acid. **Australian Journal of Plant Physiology**, v. 26, n. 4, p. 387-390, 1999.

PESIS, E.; FUCHS, Y.; ZAUBERMAN, G. Starch content and amylase activity in avocado fruit pulp. **J. Amer. Soc. Hort.** Israel, v. 103, n. 5, p. 673-676, 1978.

PETERSON, G. L. A simplification of the protein assay method of Lowry et al. which is more generally applicable. **Anal. Biochem.,** Baltimore, v. 83, p. 346-356, 1977.

PIJLS, L. T. G.; TIMMER, A. A. m; WOLDEGEBRIEL, Z. Cultivation, preparation and consumption of enset (ensete vitricosun) in Ethiopia. **J. Sei. Food Agric.,** London, v. 49, (2): p. 107-111, 1995.

PRABHA, T. N. and BHAGYALAKSHMI, N. Carbohydrate metabolism in ripening banana fruit. **Phytochemistry,** Kiel, v. 48 (6): p. 915-919, 1997.

PURGATTO, E.; LAJOLO, F. M; NASCIMENTO, J. R. O.; CORDENUNSI, B. R. Inhibition of 0-amylase activity, starch degradation and sucrose formation by IAA during banana ripening. **Planta,** Berlin, v. 212 (5-6): p. 823-828, 2001.

ROGERS, S. W., ROGERS, J. C. Cloning and characterisation of a gibberellin- induced Rnase expressed in barley aleurone cells. **Plant Phisiol.,** Bethesda, v.119, p.1457-1464, 1999.

SAMBROOK, J.; FRITSCH, E. F.; MANIATIS, T. Detection and Analysis of Proteins

Expressed from Cloned Genes. In: SAMBROOK, J.; FRITSCH, E. F.; MANIATIS, T . **Molecular Cloning: laboratory manual, v.** 3, chap. 18, p. 75; 1989

SEYMOUR G. B. Banana. In: SEYMOUR G. B" Taylor J. E" Tucker G. A. *Biochemistry of Fruit Ripening,* London: Chapman & Hall .2ed. ,p. 83-101, 1996.

SONNEWALD, V.; QUICK, W. P.; MACRAE, E.; KRAUSE, K. P.; STITT, M. Purification, cloning and expression of sucrose phosphate synthase in *E. coli.* **Planta,** Berlin, v. 189, p. 174-181, 1993.

SPEIRS, J., BRADY, C. J. Modification of gene expression in ripening fruit. **Aust. J. Plant Physiol**..Melbourne, v. 18, p. 519-32, 1991.

STEUP, M. Starch degradation. In: Sumpf, P. K and Conn E. E. (eds). *Carbohydrates, in The biochemistry of plants. A comprehensive treatise*, San Diego, California. Academic Press. V. 14, p. 255-296, 1988.

THORNTON, T. M.; SWAIN, S. M.; OLSZEWSKI, N. E. Gibberellin signal transduction presents ... the SPY who O-GicNAc"D me. **Trends in Plant Science,** v. 4(11), p. 424-428, 1999.

TUCKER, G. A., GRIERSON, D. Fruit ripening. In: Daves, D.D., ed. **The Biochemistry of Plants:** A comprehensive treatise. San Diego: Academic Press, v.12, p. 265-318, 1984.

VENDRELL, M. Reversion of senescence: effects of 2,4-dicholorophenoxyacetic and indoleacetic acid on respiration, ethylene production, and ripening of banana fruit slices. **Aust. J. biol. Sei.,** v. 22, p. 601-610, 1969.

VENDRELL, M. Acceleration and delay of ripening in banana fruit tissue by gibberellic acid. **Aust. J. biol. Sei.,** v. 23, p. 553-559, 1970.

URIO, M.; EIGEL, W. E. Duration of kilning treatment on α-amylase, β-amylase and endo (1,3) (1,4)-β-D-glucanase activity of malted sorghum *(Sorghum bicolor).* **Process Biochemistry,** v. 35, p. 433-436, 1999.

URIO, M.G. Changes in enzyme activies during germination of cowpeas (Vigna unguiculata, cv. California blackeye). **Food Chem.,** v. 73, n. 1, p. 7-10, 2001.

WANG, F.; SANZ, A.; BRENNER, M. L.; SMITH, A. Sucrose synthase, starch accumulation, and tomato sink strength. **Plant Physiol.,** Bethesda, v. 101, p. 321 - 327, 1993.

WANG, S. M.; LUE, W. L.; EIMERT, K.; CHEN, J. C. A phytohormone-regulated beta-amylase gene expression in rice. **Plant Mol. Biol.,** Belgium, v.31, n. 5, p. 975982, 1996.

WEGRZYN, T.; MACRAE, E. Alpha-amylase and starch degradation in kiwifruit. J. **Plant Physiol,** Bethesda ,v. 147, p. 19-28, 1995.

WHITAKER, J. R. **Principies of enzymology for the food sciences.** 2ed., New York, Mareei Dekker, 1994, 625p.

WONG, D. W. S. Amylolitic Enzymes in **Food Enzymes: Structure** and **Mechanism**, chap. 3, p. 37, ed. Chapman and Hall, USA, 1995. 390p.
YANG, S.F. Regulation of ethylene biosynthesis. **HortSci.,** v. 15, p. 238-43, 1980.

YOSHIOKA, K. Changes in saccharides and enzyme-activities related to carbohydrate-metabolism during growth and development of fig (F/cus *carica* L) fruit on the tree **Journal of the Japanese Society for Food Science and Technology-Nippon Shokuhin Kagaku Kogaku Kaiashi** V. 42, n. 4, p. 219-224, 1995.

YOUNG, R.E., SALMINEN, S., and SORNSRMCHAI, P. Enzyme regulation associated with ripening in banana fruit. **Coll. Int. Centre National Recherche Scientifique,** N. 238, p. 271, 1974

YU, S. M.; KUO, Y. H.; SHEU, G.; SHEU, Y. J.; LIU, L. F. Metabolic depression of a-amylase gene expression in suspension-cultured cells of rice. **The Journal of Biological Chemistry,** v. 266, n. 31, p. 21131-21137, 1991.

YU, S. M.; LEE, Y. C.; FANG, S. C.; CHAN, M. T.; HWA, S. F.; LIU, L. F. Sugars act as signal molecules and osmotica to regulate the expression of a-amylase genes and metabolic activities in germinating cereal grains. **Plant Mol. Biol.,** Belgium, v. 30, p. 1277- 1289, 1996.

ZABADAL, T.G.; DITTMER, T. W. Gibberellic acid spray increase berry size and reduce shot berry of "Vanessa" grapevinis. **Journal America Pomological Society,** v. 54, n. 3, p. 130-133, 2000.

SUMMARY

Ripening is an exclusive stage of fruit development, involving a series of metabolic transformations from different energy sources. It is mediated by a dynamic enzyme complex, resulting in the synthesis/degradation and conversion of compounds that will make the fruit acceptable for consumption. Depending on the type of fruit, this energy source can be in the form of organic acids, sucrose from the plant itself or in the form of starch. The banana *(Musa acuminata* L.) is a climacteric fruit that uses starch as its main carbon source, which is reduced during the climacteric period from levels ranging from 12 to 20% to less than 1%. In addition to this degradation, the sucrose content can reach up to 15%, depending on the cultivar. Gibberellic acid (GA_3) is a phytohormone from the gibberellin family that has been widely studied in cereals because it increases gene transcription of a-amylases. In fruit, it is responsible for maintaining firm texture, soluble solids content and delaying ripening. When studying the influence of gibberellic acid on the starch-sucrose metabolism in banana slices, it was observed in this study that the phytohormone did not alter the respiratory peak or ethylene synthesis. However, it delayed the degradation of starch and the accumulation of soluble sugars for three days. This delay was accompanied by a decrease/delay in the activity of enzymes that degrade starch and synthesise sucrose, a- and p-amylases and sucrose phosphate synthase, respectively, and no difference was observed in the increase in gene expression of sucrose phosphate synthase and phosphorylases.

MIX
Papier aus verantwortungsvollen Quellen
Paper from responsible sources
FSC® C105338

FSC
www.fsc.org

Printed by Books on Demand GmbH, Norderstedt / Germany